TACTICAL TRACKING OPERATIONS

The Essential Guide for Military and Police Trackers

David Scott-Donelan

TACTICAL TRACKING OPERATIONS

The Essential Guide for Military and Police Trackers

David Scott-Donelan

Paladin Press • Boulder, Colorado

Tactical Tracking Operations: The Essential Guide for Military and Police Trackers
by David Scott-Donelan

Copyright © 1998 by David Scott-Donelan

ISBN 1-58160-003-8
Printed in the United States of America

Published by Paladin Press, a division of
Paladin Enterprises, Inc., P.O. Box 1307,
Boulder, Colorado 80306, USA.
(303) 443-7250

Direct inquiries and/or orders to the above address.

PALADIN, PALADIN PRESS, and the "horse head" design
are trademarks belonging to Paladin Enterprises and
registered in United States Patent and Trademark Office.

All rights reserved. Except for use in a review, no
portion of this book may be reproduced in any form
without the express written permission of the publisher.

Neither the author nor the publisher assumes
any responsibility for the use or misuse of
information contained in this book.

Visit our Web site at www.paladin-press.com

CONTENTS

Glossary of Tracking Terms		xiii
1.	Tactical Tracking: An Introduction and History	1
2.	The Making of a Tracker: Skills and Attributes	13
3.	Tracking Techniques	29
4.	Conducting the Follow-Up	55
5.	Team Tracking and Tactics	61
6.	Alternative Tracking and Follow-Up Methods	75
7.	Countering the Tracker	85
8.	Command and Control of a Tracking Team	107
9.	High-Tech Help for Tracking Teams	121
10.	Development of a Tracking Team	129
11.	Training the Tracker	135
12.	Weapons and Equipment	151

DISCLAIMER

The author has tried to provide professional, accurate, helpful data, recommendations, procedures, and suggestions. However, no responsibility whatsoever is assumed by the author or publisher for any errors of omission or commission or for any matter, or result based on any matter, item, or other information presented in this book. Such responsibility belongs to and remains solely with the individual reader using or acting upon any of the above. This book is presented for academic study only.

This book is dedicated to the memories of:

Lieutenant Robin Hughes, MFC

Sergeant Russell Williams

Sergeant Nick Edwards

Sergeant Richie Smith

of the **Rhodesian Tracker Combat Unit**

and

Sergeant Peter White, BCR

Corporal Douglas Cookson

of the **Rhodesian Light Infantry**

ABOUT THE AUTHOR

David Scott-Donelan was a career soldier spanning almost three decades of active duty in the war zones of Rhodesia, South Africa, Mozambique, and South-West Africa/Namibia.

Enlisting in the Army of the Federation of Rhodesia and Nyasaland in 1961, Scott-Donelan was one of the original members of the resuscitated C Squadron (Rhodesia) Special Air Service, where he was introduced to the concepts of irregular warfare and tactical tracking by Allan Savory, a game ranger known for his innovative and successful concepts in hunting down heavily armed elephant and rhino poachers.

In 1968 Scott-Donelan was posted to the new Tracker Combat Unit (TCU), commanded by Allan Savory, with the mission of tracking down and annihilating Communist-trained and equipped nationalist insurgents infiltrating the Rhodesian border from Zambia and Mozambique. He went on to command the TCU and was responsible for the selection and training of expert trackers for the unit, which was beginning to make a name for itself on operations. In 1974 the TCU was absorbed by an innovative new counter-insurgency unit known as the Selous Scouts, and Scott-Donelan was posted to the Rhodesian Light Infantry (RLI), which was heavily involved in helicopter and airborne operations against armed terrorist gangs infiltrating Rhodesia in increasing numbers. After several years of nonstop action in the RLI, he served as an intelligence officer at a Brigade HQ and Combined Operations HQ, Rhodesia's equivalent of the Pentagon. Frustrated with staff duties, he agitated for a transfer to the Selous Scouts and was appointed Officer Commanding Training Group, which included the Tracking and Bush Survival School, the notorious "Wafa Wafa," on the shores of Lake Kariba.

In 1980, due to intense political pressure from the United States, Britain and the United Nations, Rhodesia, after having never lost a battle, lost the war and became the Republic of Zimbabwe.

Joining the South African Special Forces in 1980 as a member of 5 Reconnaissance Regiment, he commanded the regiment's Developmental Wing, which was responsible for establishing a complete training and operational resource base as well as conducting training programs for several guerrilla armies. Five years later he was seconded to the South-West Africa Territorial Force as a company commander and made responsible for operations against the Peoples Liberation Army of Namibia infiltrating into South-West Africa/Namibia from Angola and Zambia.

Immigrating to the United States in 1989, he is now the training director of the Tactical Tracking Operations School, which trains law enforcement, corrections, and military personnel in the same tracking techniques that proved so successful against armed and dangerous fugitives in Africa.

PREFACE

This book is not the work of an armchair theorist. It is based on more than 25 years of active counterinsurgency tracking experience from both an operational and training point of view. Having done my fair share of tracking armed and dangerous fugitives, this book reflects some of my experiences on the trail as well as with the planning and conducting of training programs. Police and corrections officers who have been trained in tactical tracking techniques asked me to write this book to draw attention to the value and usefulness of tactical tracking to the law enforcement community and also explain how a police department or corrections institution can go about establishing and training a tracking team of its own.

It has been said that tracking is a dying art, and I suppose in some ways it's true. Many Native American Indians, once masters of "tracking and trailing," have lost these skills, and even the U.S. Border Patrol has virtually abandoned its man-tracking activities in favor of high-tech detection sensors and unassailable border barriers. All of this prompts the question: does the ancient skill of tracking have a place in law enforcement activities today?

There are those who will disagree, but I say yes there is. There is a place for trained trackers not only in the search and rescue role but in tracking down and recovering armed fugitives, escaped convicts, and fleeing felons, who all pose a threat to the safety of our communities.

"OK," you say, "but how and where can we get the necessary training to do this?"

There are a number of people in the United States who teach a two- or three-day search and rescue tracking courses, and some of them are very good. I know of one community college that provides several two-day tracking courses to law enforcement officers each year. Jack Kearney, the master tracker from the U.S. Border Patrol, has retired, but his excellent book, *Tracking: A Blueprint for Learning How*, is still available from several sources. You can learn a great deal from Jack's book, and I recommend it wholeheartedly. The Tom Brown School, situated in the Pine Barrens of New Jersey, teaches tracking to the public for a fee and also publishes several good books on animal tracking techniques, which is a good way to start. Many outdoor and hunting magazines contain articles and tips on tracking animals. So one way or another, there are ways to become familiar with the primitive tracking skills that were so essential to civilizing mankind.

The astute reader will notice from the above list of training sources that there is no mention of "man tracking." Lately, tracking training in the United States has been focused more on search and rescue operations than on fugitive recovery. This book was written to fill the gap, because at no other time in history has the rural and wilderness areas of this country been faced with such a wide range of criminal activities as they do today. Renegade militia groups, antisocial survivalists, criminal gangs, marijuana cultivators, fanatical religious cults, environmental saboteurs, toxic waste violators, timber thieves, moonshine distillers, and wildlife poachers all hang out in the backcountry areas of the U.S. and pose a constant threat to the peace and well-being of the nation.

Even if tracking is a dying art, it is not dead yet. Technology does play a major part in crime detection, but nothing can replace the human faculties of sight, hearing, smell, touch, and intuition, all of which are put to maximum use when tracking a fugitive. These traits are better developed in some than others, but they can be trained and honed. No, tracking need not be a dying art. Although it has been placed on the shelf and temporarily forgotten, the time has come to bring it back and put it to work for the benefit of the peace, safety, and welfare of our families, communities, and country.

—David Scott-Donelan

e-mail address: mantrack@aol.com
Web site: http://members.aol.com/mantrack/

GLOSSARY OF TRACKING TERMS

When commencing any new activity, it is inevitable that there will be unfamiliar words, terms, and expressions that will have to be learned and understood by a newcomer. So it is with tracking. Although we have done our best to keep technical jargon to a minimum, there are a few words and phrases that cannot be simplified. This brief glossary of tracking terms is included to explain them to the reader.

Spoor: Spoor is *a set of tracks laid on the ground* that are visible to a tracker. For example, "following the spoor." Spoor is totally interchangeable with the words tracks, trail, and set of prints.

Follow-up: A follow-up is *the physical act of following a set of tracks* on the ground made by a fugitive. A follow-up is conducted by either a single tracker or a team of trackers. For example, "The follow-up commenced at first light."

Tracking team: When tracking armed and dangerous fugitives, a four-man tracking team is used. It is *a self-contained tactical unit* that provides its own protection and moves in various formations according to the terrain and vegetation conditions.

Tracker: The term tracker refers to *the member of a tracking team who is physically looking for and following the tracks* or more generally somebody who is a member of a tracking team or *a person who is able to follow a set of tracks*.

Controller: The controller is *the person who controls a follow-up* and is responsible for its tactical movement and formations. He moves behind and protects the tracker.

Flank tracker: There are two flank trackers in the four-man tracking team positioned on each side and slightly ahead of the tracker. Their job is to *protect the tracker and controller* from ambush and to assist in the search for lost tracks.

Three-sixty (360): When involved in lost spoor procedures, trackers move in a circle in an attempt to relocate the tracks. This circle is referred to as a three-sixty, after the 360 degrees in a circle.

Guerrilla: In this book, the word guerrilla is interchangeable with fugitive, felon, terrorist, insurgent, or infiltrator and refers to an armed criminal.

Quarry: This word is used as an alternative to guerrilla, fugitive, or "the pursued."

Chapter 1
TACTICAL TRACKING: AN INTRODUCTION AND HISTORY

There is no hunting like the hunting of man, and those who have hunted armed man long enough never care for anything else thereafter.

Ernest Hemingway

This book was originally conceived as a training resource for military special operations units and personnel involved in low-intensity counterguerrilla warfare and not specifically for the law-enforcement and corrections communities. However, the situation in the United States has reached a point where the gap between military and police tactical operations has narrowed almost to a point where it no longer exists. Camouflage uniforms, helmets, body armor, automatic weapons, sniper rifles, grenades, and assault tactics are now common to police departments all across the country. Indeed, it would be hard to tell the difference between a SWAT and a military special operations team by appearance alone. Much of what soldiers do in war is now duplicated by police officers in peacetime except that the average tactical officer in one of our larger cities spends far more time on the urban battlefield than any of his military counterparts.

Just as fire and maneuver have become part and parcel of a tactical police officer's repertoire, there are times when a tracking capability will be required by special response teams (SRTs) from both police and correctional departments. This book is intended to address this need, and although it was written by a soldier based on his 25 years of continual irregular warfare operations, police and correctional officers will find much that is relevant and applicable to their specific type of operations.

Tactically speaking, there is no difference between a military combat tracker team following up a band of terrorists through an Asian jungle and a correctional SRT tracking a group of escaped prisoners through the pine forests of the Pacific Northwest or a sheriff department's SRT hunting armed marijuana growers in the Californian foothills. To illustrate this point further, trackers of the Inmate Recovery Team from the Washington State Penitentiary at Walla Walla have already conducted successful follow-ups in the farm country surrounding the jail and recovered several prisoners who had managed to escape. Prior to the creation of this team, escaped convicts, unless they were apprehended in the immediate vicinity of the jail, would be left for local or state law enforcement to deal with.

Most police officers who have received tracking training have been surprised at how often their newly acquired skills have aided them in their everyday activities. To support this statement, let me quote from comments made by officers who attended the first operational tracking course ever run in the U.S., sponsored by the Washington State Criminal Justice Training Commission in April 1994.

As a K9 handler from a rural county, this training is a must. The knowledge and skills learned will greatly enhance the proficiency and success of a K9 team as well strengthen officer safety skills while on the track. After completing this training the dog handler becomes a greater asset to his four legged partner.
    Deputy, Grays Harbor County
    Sheriff K9 Division.

Excellent training that will be applied to the recovery of escapees from Clallam Bay Corrections Facility as well as to strengthen my skills in my search and rescue role.
    Sergeant, Clallam County Sheriffs'
    Search and Rescue Team

This type of training is the most effective way I know of finding bad guys who run from a crime scene on foot.
    Detective and SRT Commander,
    Grays Harbor, WA, Sheriff's Dept.

As a K9 officer, tracking training will help me on a daily basis. This course has opened a new awareness of K9 handling.
    K9 Officer, Centralia, WA, Police
    Dept.

Tracking skills will help my department to develop new means of apprehending escapees and for getting a better indication of avenues used for escape.
    Investigator, Dept. of Corrections,
    Washington State Penitentiary

Tracking is a necessary tool for the "cop in the woods." Without the ability to identify and read spoor left by thieves, the timber cop's duties become more difficult. Unlike street cops in the cities, timber cops do not have the luxury of witnesses and must utilize all the skills they possess to ensure convictions.
    Investigator, Dept. of Natural
    Resources, State of Washington

I am sure you will agree that the above comments from working cops prove the point that tracking skills are a necessity for law enforcement officers deployed in rural areas. A word of warning, though. Tracking is a strenuous and mentally stressful activity and could be, in some cases, hazardous to your health! So be prepared.

Tracking, simply put, is a reactive effort to close with and apprehend a fleeing quarry, whether terrorist, escaped criminal, or illegal border crosser who attempts to outrun and outwit the forces of law and order. Therefore, trackers must possess high standards of endurance and concentration to close the time and distance gaps between the pursuer and the pursued. To successfully close this gap requires the ability to go on regardless of the difficulties of ground, weather, time, or countertracking activities used by the quarry. The only way to overcome these problems is to move quicker, rest less, and plan ahead so that the time and distance advantages held by the fugitive are steadily whittled down to nothing. This can only be achieved by a continuously high level of fitness, endurance, and determination. Anything less will result in failure and the escape of the fugitive.

To illustrate the dedication required to close the gap and make contact with a fugitive, consider Sgt. Joe Conway and his tracking team from the Rhodesian Special Air Service. They once tracked a gang of terrorists who had murdered a farmer for six days over rugged and broken country in temperatures often in excess of 100 degrees. Sleeping on the tracks, the team followed the gang for close to

100 miles, eventually making contact, killing three and capturing the remainder. At their trial (all Rhodesian terrorists were treated as common criminals), the gang leader complained to the court that they had been "tracked down like dogs." Quite so. This totally focused mind-set is the only way a tracker can achieve success in apprehending man's most deadly prey, man himself.

A BRIEF HISTORY OF OPERATIONAL TRACKING

Man by nature is a predator, and his ability to track and kill animals was probably the most important factor determining his survival as a species in prehistoric times. Without the ability to follow and eventually kill animals for food, early man would have used all his energy and time foraging for low-calorie vegetable foodstuffs. By killing animals and eating an abundance of highly nutritious, protein-rich fat and meat, man's mental and creative abilities exploded, transforming him from a low-functioning subhuman to the thinking, talking, communicating ancestor of modern man.

Colonial America

There is no doubt that tracking has played a role in many prehistoric battles and conflicts, but it is only within the past 250 years that it has played a significant part in modern warfare. The earliest records of combat tracking in America go back to the early 1700s when British settlers fought their French counterparts in seeking to colonize and dominate what eventually became the 13 original states. The skill was most likely passed on to the settlers by the local Indian tribes, and both the French and English used the Indians themselves as trackers. During this period, the British formed several militia-style units known as Provincials, with manpower drawn from the American-born descendants of the original settler families. These tough woodsmen and farmers, seasoned by years of combat against hostile Indians, became the ears and eyes of the British and performed many successful long-range reconnaissance and raiding missions against the French forces.

Probably the best known of these Provincial companies was the one commanded by New Hampshire native Maj. Robert Rogers, which became known as "Rogers' Rangers." In his famous Standing Orders written in 1759, Rogers instructed his command:

"If we strike swamps or soft ground, we spread out abreast, so it's hard to track us."

"When we march, we keep moving till dark, so as to give the enemy the least possible chance at us."

Perhaps even more revealing of the French army's ability to track, Rogers goes on to say:

"If somebody's trailing you, make a circle and come back on your tracks, and ambush the folks that aim to ambush you."

This appears to be the first reference to combat tracking in American history. Mainly due to the success of Rogers and his Rangers, tracking continued to be used with great effect during the westward expansion into the Indian territories and on to the Pacific Ocean.

The Indian Wars

In the United States of the nineteenth century, the scope and range of horse-mounted operations fostered the use of tracking skills on both sides during the Plains Wars. Although several Indian tribes produced outstanding trackers (it was called "trailing" then; the word "tracking" came into general use at a later date), particularly noteworthy were the Apaches of Arizona, New Mexico, and the Mexico border region. Brig. Gen. George Crook was the first military commander to recognize the value of Indian scouts, saying, "The best way to pursue Indians was with other Indians." General Crook had no trouble getting recruits because many Apaches were willing to enlist just to settle old scores against other bands of Indians. So well did these scouts perform with both trailing and reconnaissance tasks that General Crook provided them with a full uniform and awarded them their own guidon complete with white crossed arrows and company letter against a red background. (It is of interest to note that during World War

II, the combined American-Canadian Special Forces unit used this emblem, and it is still used by the U.S. Army Special Forces to this day).

There is no doubt that these Apache scouts performed with distinction. An October 1902 article in the *Saturday Globe* of Albany, New York, recorded their prowess:

TRAILERS THAT ARE KEENER THAN BLOODHOUNDS—THE SCOUTS WHO LED LAWTON'S MEN IN THEIR PURSUIT OF GERONIMO

In Northern Mexico, New Mexico and Arizona there are men who far surpass the blood hound when it comes to trailing. Men who served in the campaign against Geronimo and the hostile Apaches, many of them expert trailers, were from day to day overwhelmed with astonishment at the almost superhuman instinct of the Mexican and Indian scouts, who in that memorable pursuit, followed the fleeing Apaches over sand deserts harder than asphalt and floors of solid rock upon whose bare surfaces the soldiers could see no trace of the horsemen who passed that way. Accompanying our command were about a hundred Indians, enlisted and used as scouts.

Talk of trailing! Why, I never until then dreamed that it was possible for human beings to do what these men did every day of the campaign. Frequently we would descend mountains, along the slope of which old scouts of my company were barely able to make out the trail of the Apaches, until we reached a valley about half or three quarters of a mile wide, the surface of which was as hard as adamant. Here we could see nothing, but the scouts ahead, on coming to such places never hesitated one minute, but struck boldly across, following the trail up the mountain side again. As we crossed these valleys and mounted once more up the sides of the high ridges we would again catch the traces of the Apaches in the softer and looser soil of the mountain side and many a time we have wondered how our Indian trailers were able to follow the track on ahead of us over the valleys, where the surface was as hard as asphalt and crossed by fresh trails of hundreds of cattle that had passed up or down the depression after the Indians.

The most remarkable case of trailing that ever came under my notice, however, occurred in 1887, when the San Carlos Apaches broke out of their reservation and went on the warpath south into Mexico. This occurred shortly after the Geronimo campaign, at a time when the War Department had discharged all the Indian scouts attached to the South-western outposts, thinking the trouble was all over and that there would be no longer any use for them. Consequently when we received a hurry order at Fort Huachuca to go in pursuit of the San Carlos Apaches we were obliged to leave without taking any of these human bloodhounds with us.

Colonel Lawton, while stirring about, ran across an old Mexican riding along on his burro. The Colonel asked him if he would be our scout and whether he felt himself capable of taking up and following the trail of the Indians. The old fellow gave a grunt of assent, two minutes later had found the trail and, to our unspeakable astonishment, was leading us almost at a run across the barren spot and up the mountain. In three hours time he brought us in sight of the Indians encamped in a hollow. We charged down upon them, but failed to make a capture, as they saw us in time to make an escape.

This old Mexican served as our trailer for the rest of this brief campaign until we finally overtook and captured our recalcitrant Indians. In all that period he never once missed the trail, notwithstanding the fact that we frequently passed over places where no sign of tracks were apparent to us. This, to my mind, was the most astonishing piece of tracking that I ever witnessed or heard of and I believe that the clever Mexican above all others was justly right to the title of human bloodhound.

Allowing for the fact that writers of the western frontier often exaggerated and romanticized military exploits (much like television today), there is no doubt that the Apache and mixed-blood inhabitants of the border areas were extremely efficient trackers,

San Carlos Apache scouts examining the ground for tracks during the arduous campaign against the Chiricahua Apaches in 1883. These scouts are wearing surplus uniforms left over from the Civil War, but when the action started they stripped down to loincloths and wore red headbands to distinguish them from hostile Indians.

A contempory lithograph of Gen. George Crook. General Crook, a brilliant and innovative soldier, saw the value of using Indian scouts as trackers and used them with great effect against hostile tribes, including Geronimo and his Chiricahua Apache band.

and without them the army would not have pacified the West as quickly and effectively as it did.

Colonial Africa

There is an interesting historical sideline to this period that involves the American tracking skills that were exported to British Colonial Africa. During the 1893 war between British pioneers and the Matabele tribe in what was then Southern Rhodesia (now Zimbabwe), a small column of 34 mounted troopers, led by a Maj. Allan Wilson, was tasked to pursue and engage the *impis* (warrior regiments) of Lobengula, Chief of the Matabele. Wilson had obtained the services of two American adventurers as scouts, and one of these, a Minnesotan by the name of Frederick Burnham, used his extensive tracking skills learned on the Mexican border to pursue the fleeing warriors. In order to cut down the time/distance gap it is reported that Burnham actually tracked at night, using his fingers to feel for the hoofprints of the cattle driven by the Matabele warriors. The story goes on to tell of the eventual destruction of Allan Wilson's heroic patrol on the banks of the Shangani River, but Burnham was one of only two combatants to escape the massacre with his life and (it is reputed) to return to America with a large amount of stolen gold!

Burnham eventually became the chief scout for the British Army during the Anglo-Boer War (1898–1902) and as a major, was awarded the Distinguished Service Order (DSO) by Prince Albert, the husband of Queen Victoria, for gallantry during many difficult and dangerous missions behind the Boer lines. Burnham surely must have been an outstanding man to be awarded the Queen's Commission, because during the Victorian era British military commissions were only awarded to nobility and the very wealthy. He became well known in Victorian circles and

A sketch drawn by Gen. Sir Robert Baden-Powell of Maj. Frederick Burnham, DSO., Chief Scout for the Imperial Forces of Lord Roberts during the Anglo-Boer War in South Africa, 1898–1902. Baden-Powell founded the Boy Scouts in 1904, inspired by the valuable scouting and messenger work done by boys during the Boer siege of the town of Mafeking during the war. There is no doubt that the skill and daring of Major Burnham's exploits played a part in formulating Baden-Powell's ideas on his future organization.

was the personal friend of such luminaries as Cecil John Rhodes, Winston Churchill, Lord Baden-Powell, founder of the Boy Scouts, U.S. President Teddy Roosevelt, and the immortal African hunter Frederick Courtney Selous. If there ever was a true American hero, Burnham must be considered a prime candidate, but unfortunately he is hardly known in his own country today. (For more information on this fascinating pioneer, Pony Express rider, scout, soldier, hunter, gold prospector, adventurer, and businessman, read his books *Scouting on Two Continents* and *Taking Chances*.)

Into the 20th Century

With the advent of mechanized warfare in the 20th century, the horse was virtually discarded as a form of military transport, and the static nature of trench warfare marked the end of the tracker and his skills so painstakingly learned over 200 years of early American history. Yet after World War II, the "winds of change" swept over the British Empire, and the period was characterized by a rising tide of native nationalism. In the far-flung colonies of Africa and Asia, the local people, emboldened by clandestine Soviet and Chinese support, formed underground political movements to throw off the colonial yoke and agitate for self-rule.

Lacking the necessary training and weaponry to counter the European colonists, these nationalist movements were compelled to fight as guerrillas from the relative safety of jungles and forests supported by the local indiginous people. To counter this threat, British military authorities came up with new jungle warfare tactics using small, highly mobile, lightly armed hunter-killer teams to penetrate guerrilla-held areas in a deadly game of cat and mouse. In this environment, tracking once again came into its own as a vital part of military operations.

Malaya, 1948–1957

During World War II, British forces armed and trained Chinese guerrillas living in its colonies to fight against the Japanese invaders. Most of these missions were carried out with great success, particularly in rubber- and tin-rich Malaya. Unfortunately for Britain, the guerrillas there turned out to be adherents of Mao Tse-tung, the Chinese Communist leader, and did not relish the idea of returning the country to their former colonial masters after the war.

Pressure built up, and in 1950 the so-called Malayan Peoples Liberation Army (MPLA)—of which there were hardly any Malayans—emerged from the jungle and struck at British interests throughout the country. Conventional military strategies proved to be useless in countering this movement, so new tactics were devised that required small, highly trained patrols of the Special Air Service (SAS) to penetrate the MPLA-dominated jungle areas. Initially using native Iban trackers from Sarawak on the island of Borneo, the patrols eventually made contact with the guerrilla bands and a war of attrition ensued. Most of the contacts were the result of ambush and painstaking follow-ups into the dark, dank depths of the triple-canopy jungle.

After almost a decade of warfare, the British finally prevailed over their Chinese adversaries, and peace was restored to a new political entity, the Malaysian Federation, comprising Malaya and three separate nations on the island of Borneo: Sabah, Brunei, and Sarawak.

Kenya, 1952–1962

Emboldened by the success of the Malayans in throwing off colonialism, other British territories, particularly in Africa, followed suit. The East African country of Kenya erupted into antisettler violence when gangs known as Mau Mau from the indigenous Kikuyu tribe attacked farms and homesteads of blacks and whites alike with acts of unspeakable violence in an attempt to challenge the colonial government and impose majority rule. (The excellent book *Something of Value* by Robert Ruark describes the way of life and the times very well.)

Reacting to the menace, the Kenya Police Special Branch created units made up of black and white police officers who employed tracking and "sting" type of operations to contact, convert, or wipe out the killer gangs. Records show that more than 35,000 people were murdered by Mau Mau before the conflict was finally resolved in 1962. (For an understanding of these operations, read *Gangs and Counter Gangs* by Gen. Frank Kitson or *The Hunt for Kimathi* by Ian Henderson.)

It is of interest that the Kenya Regiment, a reserve military unit composed of white Kenyan settlers, was trained by the Rhodesian Army in Salisbury during this period, and Rhodesians were involved in the conflict in Kenya itself as advisors and operations staff. This factor was to play an important part in future counter-guerrilla operations in Southern Africa.

Southeast Asia, 1962–1966

Meanwhile back in Asia, President Sukarno of Indonesia, having thrown out the Dutch colonial regime, was secretly making plans to invade the three Malaysian Federation countries situated on the northwestern side of the mostly Indonesian island of Kalimantan (formerly Borneo). Sukarno, an ambitious leader of grandiose style, was eager to create a supernation called "Maphilindo" that would embrace the new Malaysian Federation, the Philippines, and the 3,000 islands that make up the Republic of Indonesia. Aware of Sukarno's expansionist plans, the British girded for war and deployed the SAS back into Southeast Asia.

On April 12, 1963, a platoon of Indonesian troops attacked a police station inside Sarawak in an attempt to test British resolve and evaluate reactions. Fighting the Indonesians at their own game, the British placed a premium on skills more suited to a hunter—the arts of concealment, endurance, tracking, local knowledge, and weapons proficiency, which were honed to perfection during the Malayan campaign. Tony Geraghty, in his book *Who Dares Wins*, describes a typical example of this tactical philosophy:

> *Turnbull . . . achieved an eye for spoor as accurate as his native tracker . . . reading the splayed toe prints of an aborigine for what they were; the terrorist's footprint, which invariably revealed cramped toes that had once known shoes; and spotting a fine human footprint imposed by the more canny walker on an elephant footmark in an attempt to blur the trace. Turnbull once followed the tracks of four men for five days, until he spotted the hut they were occupying. He then waited for an impending rainstorm to arrive, correctly guessing that the sentries would take shelter, and drew to within five yards of the hut before killing the four guerrillas. According to one officer who served with him, Turnbull used a repeater shotgun with such speed and accuracy that "it would fill a man with holes like a Swiss cheese."*

The British and their Commonwealth allies eventually prevailed, although the conflict, virtually unknown by the rest of the world until recently, continued for three years before the Indonesians capitulated. With it, the dream of Maphilindo died. (For more information on this conflict, read *Britain's Secret War in South East Asia* by Peter Dickens.)

Southern Africa-Rhodesia, 1961–1980

Attempts to overthrow other colonial regimes throughout the Third World were not lost on the military leaders of what was to become Rhodesia. The self-declared Republic of Rhodesia of 1965 (now Zimbabwe, 1980) resulted from the political breakup of the Federation of Rhodesia and Nyasaland, which was a futile attempt within the British Commonwealth to deal with the problems of the three territories that were supposed to constitute the federation: Southern Rhodesia became Rhodesia, Northern Rhodesia became Zambia (1964), and Nyasaland became Malawi (1964).

In 1961, the Rhodesian SAS—which had been raised and dispatched to reinforce the British in Malaya in the 1950s and disbanded after that conflict—was resurrected together with a light infantry battalion operating along commando lines, an armored car squadron, and several squadrons of aircraft suitable for counterinsurgency operations. Rhodesians who had operated with the British in both Malaya and Kenya were now in senior positions in the army, including Gen. Peter Walls, who had commanded C Squadron, Rhodesian SAS, as a major in Malaya.

In 1962, Allan Savory, a Rhodesian game ranger turned ecologist who had successfully used tracking techniques to hunt elephant poachers, presented his tracking operational concepts to the Rhodesian SAS, which were very well received. However because of the political breakup of the Federation of Rhodesia and Nyasaland, his plans were put on hold until 1965, when selected members of the Rhodesian SAS commenced training in the Rhodesia-Mozambique border area. Under Savory's tutelage and with tactical input from the Rhodesian SAS, new and effective tracking team tactics were devised that transformed individual tracking skills into a deadly antiterrorist art.

Rhodesia did not have long to wait, because in late 1966 a group of 110 guerrillas infiltrated across the northern Zambezi River border from Zambia. (Ironically, the indications of the presence of this group were discovered by a game ranger who was later to become a combat tracker himself.) Rhodesian operational preparations proved to be effective: of the original 110 Communist-trained insurgents who had infiltrated the Zambezi Valley, more than 100 were accounted for, mainly by tracking teams that aggressively pursued, ambushed, and overcame the gang with speed and violent action.

Savory's wisdom and foresight paid off handsomely, and Rhodesian Army HQ ordered him to raise and train his own unit, to be known as the Tracker Combat Unit (TCU). Composed of game rangers, professional hunters, and ex-soldiers, this unit was probably the first in history whose primary mission was an antiterrorist tracking role.

After a few more halfhearted terrorist incursions into Rhodesia between 1967 and 1971 that were dealt with speedily and effectively, the war began in earnest in 1972 when large gangs of guerrillas crossed the remote northeastern border from the Portuguese colony of Mozambique, itself afflicted by a deadly Communist-inspired insurrection. Among the first units to be mobilized against this new threat was the TCU, which along with regular army units, performed with distinction.

Unknown at that time by anyone except the military high command, plans were already

A Rhodesian SAS tracker examines equipment in a guerrilla base camp concealed in the thick brush of the Zambezi Valley separating Rhodesia and Zambia. He is lightly clad, enabling him to cope with the 100+ degrees temperature. He carries a Browning semiauto 12-gauge shotgun, ideal for penetrating the thick jesse bush found throughout the valley.

Although this was kept top secret at the time, Rhodesian tracking instructors trained elements of the Portuguese army in both Angola and Mozambique. Known as Pisteros de Combat, Portuguese trackers conducted themselves well against Marxist guerrillas in both colonies. Seen here posing with Portuguese army trainees, Capt. Brian Robinson (center rear) and the two bare-chested soldiers, all from the Rhodesian SAS, were instrumental in establishing tactical tracking as a successful counterguerrilla activity.

being made to counter the new guerrilla offensive. The insurgents had, in the Maoist tradition, infiltrated the local population, so a new antiterrorist unit was being formed that would provide intelligence on guerrilla strengths and movement within the country. This unit, known as the Selous Scouts (named after the African hunter Frederick Courtney Selous), was based on a successful irregular Portuguese unit made up of ex-FRELIMO guerrillas, known as *flechas* (arrows) and run by a legendary professional hunter, Daniel Roxo.

(In the years cited above, the then Marxist FRELIMO [The Mozambique Liberation Front] was the political and military organization fighting for independence of the then Portuguese colony. FRELIMO is to this day the paramount political organization in Mozambique, which became independent in 1974 after the radical change of government in Portugal.)

The Selous Scouts absorbed the members of the TCU and eventually reached a strength of close to 1,000 specially selected men, all parachute trained and skilled in counterguerrilla tracking operations. This unit, which also performed sting and raiding operations, was eventually credited with over 70 percent of all guerrilla casualties throughout the ensuing eight years of escalating conflict. The TCU continued to operate a bush warfare and tracking school and trained hundreds of soldiers in the skills advocated by Allan Savory back in 1962.

The founding commanding officer of the Selous Scouts, Lt. Col. Ron Reid-Daly, had also served with Gen. Peter Walls' SAS squadron in Malaya and believed that his experiences there had given him an excellent foundation in the type of warfare then facing Rhodesia. General Walls later spoke of the many lessons learned by the original Rhodesian SAS in the jungles of Malaya and passed on to new generations of Rhodesian soldiers: "It is fair to say that many of the successes of the Rhodesian war had their beginnings in Malaya. We learned what it was like to be ambushed; what the principles were in establishing our own ambushes; what gives you away; and learned the technique of tracking."

The British military historian John Keegan, in his book *World Armies*, wrote: "The Rhodesians waged a campaign of extreme military professionalism that will deserve a place in the World's Staff College courses for many years to come."

Rhodesia was eventually forced to capitulate to majority rule in 1980 due to intense political pressure from Great Britain and the United States and United Nations trade sanctions. The tracking torch, however, had already been passed to South Africa, where the basic skills and tactics that proved so effective in Rhodesia were modified to suit local terrain and conditions with great success. Many Rhodesian soldiers and police officers relocated to South Africa, taking with them their hard-won skills and operational experience.

In 1993, a group of U.S. Army Green Berets of the 5th Special Forces Group was dispatched to Zimbabwe to attend the same school where the bulk of Rhodesia's trackers were originally trained. So combat tracking, originating in America 250 years ago, exported to Africa and Asia, was finally back home where it first began.

Truly the wheel has turned full circle.

Chapter 2

THE MAKING OF A TRACKER: SKILLS AND ATTRIBUTES

It takes less time to learn a subject properly for the first time than to correct a learned mistake later.

Steve Mattoon,
Senior Tactical Instructor for
Defense Technologies

Tracking is not magic. It is a set of human skills used in an aggressive and motivated manner for the purpose of covertly following and recovering a fugitive, or fugitives, from justice. Tracking skills are basic to our primitive past and are alive and well in our collective memory banks. All we have to do is reach inward and retrieve the data!

A tracker needs a multitude of skills that when exercised in concert ensure optimal tracking performance. He needs to be able to follow, interpret, anticipate, and, most important of all, *react*. Should one of these critical skills be absent, the whole performance will be downgraded accordingly. Remember that in tactical tracking, failure can range from the inability to apprehend your quarry to sudden death in an ambush.

Like every other worthwhile activity, tracking skills need to be honed to near perfection if good performance is to result. However, we must accept the fact that some people will never make the grade no matter how hard they try. If this were not a fact of life, we would have a nation of NBA stars, all earning $10 million a year! Deficiencies in tracking skills can be worked on, but not everyone will end up a star.

There are four tracking roles—the controller, the tracker, and two flank trackers—each of which requires slightly different attributes. So even if you don't shine as a tracker, you may well be a first-rate controller, a position that calls for more tactical skill and planning ability.

Reading this book will not make you a tracker, so don't expect it to. After all, you cannot expect to be able to play a violin just by reading a book. But what this book can do is show you the way if you are a beginner, give you a new perspective if you have already mastered the art, and explain exactly what a tracking team can do for your jurisdiction and your community.

ESSENTIAL PHYSICAL AND MENTAL SKILLS

The following physical skills and mental attributes are essential for anyone who aspires to be a first-rate tracker.

Fitness

Tracking is an extremely arduous activity. Your quarry may have crossed open desert in the cool of the night; you may have to cross it in the heat of the day. Your quarry may choose to rest, but if you wish to close the gap (known as the time/distance gap), you cannot afford to rest for as long, if at all.

Ideally, as a tracker you should be able to stay on tracks for at least 48 hours carrying all your essential operational gear on your back. This level of activity mandates extreme fitness—anything else is just not good enough. An unfit person tires easily and loses concentration and mental awareness, which could result in danger to himself and others. Therefore, a tracker should be the type of person who can easily run 5 miles or walk 10 miles at least three times a week. Once a follow-up commences, the tracker should relentlessly focus on his quarry. Nothing—be it terrain, inclement weather, or any other obstruction—should divert him from his task. Anything less than peak fitness is just not good enough.

Visual Acuity

Visual acuity is more than just a fancy way of saying good eyesight. It is the ability to see and to interpret what is seen so that intelligent assumptions about a fugitive's intentions can be made.

Everybody is different, and so it is with eyesight. Urban folk usually do not have the acute vision enjoyed by country folk, mainly because the urban world is enclosed by buildings, walls, and other barriers. Eyesight, in a city, although of 20/20 vision, never gets to stretch to maximum focal power. Eyes have muscles too, and like all muscles in the human body, they have to be exercised to achieve ultimate performance. If not exercised, they tend to atrophy and lose staying power. Any muscle weakness can be improved by the correct training exercises, and trackers should always make opportunities to exercise and train their eyesight. It cannot be over-emphasized that 100 percent vision is mandatory as far as trackers are concerned.

There are, however, several types of vision deficiency that cannot be corrected and that should disqualify those afflicted from the tracking task.

Color Blindness

Color blind people, although having 20/20 vision, may miss essential clues in what is usually a green-rich environment. Most color blind people cannot tell the difference between red and green, and this inability may result in being unable to see blood spoor, for example. The following story illustrates how color blindness can have important consequences to the success of an operation, in this case a commercial venture.

A friend of mine was the general manager of a chrome mine in Africa. It produced some of the highest grade chromite in the world that was highly sought after by metallurgical companies. Extracting the ore, however, was not an easy task due to the presence of pockets of a dense, tough, green material in the ore body.

Like all good businessmen, my friend sought professional advice, and a geologist duly arrived to examine the problem. After several days of investigation, the geologist informed my friend that the green material, although troublesome, was in fact a blessing in disguise. The rock turned out to be a variety of jade known as nephrite and was worth far more than the chromite in which it was found.

Capitalizing on his good fortune, my friend decided to hire a large labor force to hand-pick through the extensive waste dumps where the troublesome green rock had been deposited. Duly hired, the local African laborers arrived for their first day of work and were briefed on what to gather from the huge piles of waste rock. Holding up a piece of the green jade, my friend explained to the crew members that this was what they had to seek and place into baskets provided. After the briefing, the new crew nodded in assent to indicate that they understood what was required of them.

Leaving them to carry on, my friend went away to attend to other duties.

At the end of the day, he went over to the dumps expecting to find a small fortune in valuable jade neatly gathered in the baskets, but he was surprised to find that all the baskets were empty. Questioning the crew, he was told that although they had searched all day, they had not been able to find a single piece of jade. Suspecting some type of industrial indiscipline, my friend angrily searched around and quickly found a fairly large piece of jade and held it out to the crew. "Do you see this?" he asked. "Yes boss," replied the crew, "we see it." My friend then handed the jade to each person in turn to make sure that everyone knew exactly what he was expected to look for and told them to return for work the next day.

The next day, my friend sat and watched his labor force scour the dump, but after several hours no jade had been collected. Calling them together, he angrily informed them that they were dismissed and to leave the property immediately.

A little later he walked over to the dump and in the space of several minutes collected a considerable amount of jade. At that moment a black thundercloud looming over the surrounding hills opened up and soaked him to the skin in a brief but substantial shower, as is common in that part of Africa. Taking off his sunglasses to wipe them, he noticed that the rain had washed the dust off the pieces of jade in the basket and they lay there gleaming in the bright sun. Suddenly a thought dawned on him, and he called over another worker who was dumping waste from a tipper truck. After telling the driver to put on the glasses, he asked him whether he could see the jade in the basket. "Yes" the driver replied, "I can." Told to take the glasses off, the driver was asked again whether he could see the green rocks. Puzzled, the driver answered in the negative.

My friend had solved the puzzle. It turned out that for some unexplained reason the people of the local tribe were color blind and could not make out the color green, which to them showed up as gray. Rushing into town, he purchased as many pairs of polarizing sunglasses as he could find and gave them to his rehired crew, who soon rewarded him with many thousands of dollars worth of carving-quality jade.

Depth Perception Problems

Poor depth perception is the second major vision problem that could disqualify someone from becoming a tracker. People suffering from a lack of depth perception tend to see things on the same plane, so they cannot make out, for example, a person hiding behind a bush or concealed in shadows. Their ability to shoot, particularly in estimating range, is also not up to the standard required of a tracker.

There is another group of people who will never make a good tracker, irrespective of their standard of eyesight. These unfortunate people, and we all know one, are those who "look" but just cannot "see." It doesn't matter how often we point things out; they just cannot see what is plainly right in front of their faces until they trip over it.

Trackers, indeed all law enforcement officers whose activities take them into backwoods or wilderness areas, must have or at least develop the ability to see *through* things. Not just to look at the trees and bushes immediately to their front or side but to see through and beyond, not limiting their vision to a restricted plane. This can be likened to the functioning of a camera zoom lens, which has the ability to focus at different distances depending on its setting.

A good way to overcome the tendency to see only within a limited plane is to go into the woods and, standing in one spot, focus your eyes on the nearest tree. Shift your gaze to the tree behind, then the next and the next until the limit of your vision is reached and all you see is a wall of trees. This will help sharpen your vision at varying ranges and reduce the amount of time it takes for your eyes to focus at each range.

This exercise can be varied by creating an

"S" type of movement with your eyes, swinging your gaze from left to right until the limit of your vision has been reached. By widening the S you will find that not only will your vision improve but that your depth perception will sharpen. You will not only begin to see correctly; your brain will start to combine what you see with what you hear and smell, thereby subconsciously interpreting and cataloging this information into safe or danger categories. This is how the elusive "sixth sense" is developed: when your brain subconsciously assesses and analyzes incoming data from your eyes, ears, and nose and compares those data inputs with previous experiences buried in your memory, often triggering a sense of danger in your conscious mind. This has happened to me on several occasions, and many people I have spoken to report the same experiences. The only common link between these people is that they are outdoors types who have developed their natural faculties to a high degree.

I must emphasize the absolutely vital necessity of good visual acuity if you are intending to become a tracker. The following story, I hope, will stress the point.

Back in the 1970s, my tracking team was called out to assist a police manhunt for a gang of armed criminals who had murdered a farmer in a rural area of Rhodesia. Two of the gang of five had been arrested at a roadblock and had revealed under interrogation that the other three had remained in the area of the murder waiting for the police activity to die down so they could slip away. I took the opportunity to question the two suspects about what type of clothing and footwear the remainder of the gang were wearing.

Because several days had elapsed since the murder, the trail from the crime scene was cold, and cattle and normal human movement had obliterated any remaining spoor. Despite the fact that the local police were of the opinion that the gang had left the area, a report came in two days later that a farm worker had seen a small fire burning in a remote area the previous night about 10 miles from the murder scene. Because it had been unseasonably cold, it was considered possible that the gang had made a fire to keep warm.

My three-man tracking team moved in quietly and commenced to scout out the suspect area. We quickly located the still-warm ashes of a fire, but there was no confirmation that it was made by the fugitive gang. We searched the area carefully for several hours and were about to call it off due to lack of evidence when one of the trackers, a game ranger by profession, found a tiny speck of red among the dried-out grass stalks close to the fire. (Color blind people would have not seen this piece of evidence.) The red speck turned out to be several small strands of wool adhering to a particularly annoying type of African grass seed. We knew from our earlier interrogation that one of the suspects was wearing red woolen socks! Bingo, instant confirmation. We widened the search area and found a fresh partial print at the base of a wooded, rocky outcrop known in Africa as a *kopje* (pronounced "kopee").

Calling for helicopter-borne troops, we sealed off the small hill, formed a sweep line, and moved in to search. In short order we made contact with the now demoralized gang. One who resisted was killed; another, attempting to escape from the hill, was gunned down and wounded by a blocking position. The third threw up his hands in surrender.

The whole operation was successful because of one man's ability to use his eyes to find confirming evidence. Had he not found this critical clue, the gang would have probably escaped, possibly to kill again. It is of interest to know that upon our return to the police base, we were informed that the entire operation was to be called off if our investigation of the fire sighting had proved negative.

Visual Defects from Age

As males get older, there is a tendency for their eyesight to deteriorate. This need not be a problem, because there is a simple one-day surgical procedure known as radial keratotomy which, by use of a laser, corrects any foreshortening of vision. However, there is nothing to prevent those people who wear glasses from becoming good trackers. I know several excellent trackers who are as blind as a bat without glasses but masters of the art with them on. In fact, if wearing specs gives you a visual edge, then by all means wear them when on the spoor. In fact, it has been noted that wearing sun glasses with polarizing lenses does assist some individuals in better differentiating varying shades of color when tracking aerial spoor in areas of thick vegetation.

Wearing glasses is not without problems while tracking armed and dangerous fugitives, however, the main one being that you could lose or break them in action. (Police officers will recall the Miami FBI agent who lost his spectacles on the way to the infamous shoot-out in that city.) Always make sure your glasses are fixed securely to your head, and always have a spare pair readily on hand.

Another problem with glasses is the risk that light reflecting from the lens could give you away to your quarry. Under optimal conditions, reflected light can be seen up to 10 miles away, so be sure to take such appropriate countermeasures as remaining in the shadows when reflection conditions are likely. Also note that glasses have the tendency to mist up in cold weather, fog, or rain, so always have demisting fluid and a quantity of dry lens cleaners on hand.

Peripheral Vision

Just as important as good eyesight is for a tracker, so is good peripheral vision. The ability to see out of the corner of your eye is important when following a trail because a tracker must keep visual contact with the rest of his team, even more so in the event of a possible ambush. Silent signals are an essential feature of team communications, particularly when creeping up on unsuspecting fugitives. Also important is the ability to see "out of the back of your head," meaning to be able to sense something or someone behind you. This is part of the sixth-sense phenomenon, and although not scientifically provable, it has been responsible for alerting many people to potentially dangerous conditions to their blind side.

Patience

It is a given that trackers must be patient people. If you are a short-fused, impatient type of person, tracking is not for you. The hard-charging "do it now" types will always, repeat always, spoil a follow-up by their needless haste to get the job done despite the absolute necessity, for example, to slowly and methodically search an area for lost spoor. Lost spoor procedures (described later in this book) can be painstaking and time consuming, involving considerable patience and close attention to detail (as the story of the red wool in the previous section illustrates). Any feeling of frustration or annoyance with the time taken to relocate lost spoor could cause trackers to overlook something small but potentially important.

One annoying distraction that can affect even the most patient of trackers is the inevitable interference from higher command. "What's happening? What's going on?" are typical of some of the irritating questions continually asked during a follow-up. This type of interference must be resisted by tracking teams, and deskchair commanders must be instructed firmly (but politely) that their constant questions can have no other effect but to disrupt the efficiency of the team and delay the follow-up.

Let me give you an example of how higher level meddling had a negative affect on what could have been a successful operation. This incident occurred while tracking the remnants of a terrorist group that had been mauled in a series of scattered firefights. The gang members had set out at a fast clip for the border of a neighboring country that was sympathetic to their cause. The area—scattered woodland with sparsely vegetated ground—was suitable for tracking, but it being monsoon season, short, sharp rainstorms had a tendency to wash out the tracks and were an ever-present problem.

HQ, impatient with the pace of the follow-up and ignorant of the way trackers operate, had dispatched a spotter plane back along the last reported direction of travel of the gang. The first we knew of this was when the plane swooped low over our position, whereupon I impolitely told the pilot to exit the area.

After his departure and everything had settled down once again, we returned to the tracks. Only 400 yards along the trail we discovered a resting place used by the exhausted gang. From the evidence of the spoor, it was apparent that they had scattered upon the sudden arrival of the plane, leaving a large amount of gear and weapons behind. While checking out the scene and reporting back to HQ, the clouds opened up and washed out all the spoor, causing us to terminate the follow-up in disgust. The result was that it took much longer to wind up the operation because we had to hunt down the gang members one by one.

The negative affect of the interference from HQ only came into perspective later when a tired and demoralized guerrilla was captured. He revealed to us that the group was asleep when the plane arrived overhead and that they had no idea they were being tracked by an aggressive and fresh tracking team supported by a platoon of first-class infantry soldiers.

Patience, displayed by both trackers and command and control alike, are prerequisites to successful tracking operations.

Aggression and Motivation

Tracking, by its very definition, means aggressive and meaningful pursuit. Its very success depends on the ability to pursue, close with, and annihilate or apprehend your quarry. A follow-up has a beginning, a middle, and an end, and the end must be pursuant with your goals. Far better to let apathy and defeatism characterize your enemy, forced to surrender due to your relentless and determined pursuit.

On a number of occasions I have been obliged to use indigenous or native trackers, some of whom were brilliant and exceedingly aggressive at their jobs. On other occasions I have used Bushmen, the original people of Southern Africa. Bushmen, by nature, are gentle and passive and don't like getting involved in other people's arguments. They are superb trackers and their bushcraft skills are legendary, but problems arise when they know they are getting close to their quarry. They pretend to "lose" the spoor and wander around seemingly confused, hoping that you will call off the follow-up or that the quarry will move out of the area. On one such follow-up, my Bushmen trackers ignored the tracks of the larger of two groups of terrorists and set off at a fast clip after the smaller group. Being skilled trackers ourselves, we soon corrected the situation and returned to the larger set of tracks.

The point is that without an aggressive spirit, it may be just as well to pack up and go home. As far as the tracker is concerned, motivation must be a driving force that sees no barrier to operational success. The strange thing about tracking operations is that better motivated people get better results, and it is the same in other walks of life too. Motivation must be a given for a tracker, and the words "I give up" should not belong in his vocabulary.

MARKSMANSHIP AND WEAPONS PROFICIENCY

Please note that both police and corrections personnel involved in tracking dangerous fugitives must be aware of and strictly adhere to their department's rules and policies concerning rules of engagement and the use of deadly force. All team members must be made aware of all rules and policies involving the use of firearms in potentially violent confrontations on an ongoing and regular basis.

It is the ultimate humiliation for a tracker to make contact with his quarry after a lengthy follow-up and then miss his shot. We will discuss suitable weapons in a later chapter, but suffice it to say it is absolutely imperative that trackers be first-class shooters, whatever the conditions, whatever the weather, whatever the position, whatever the light conditions, whatever the circumstances, and no matter what weapon they are armed with.

First and foremost, *trackers should like shooting*. This may sound strange to anybody who has purchased this book, but I am constantly amazed how many police and corrections officers I meet who just do not enjoy shooting. And the excuses they make! "I'll only have to clean my gun afterwards," "The noise bothers me," or "I shot last week." Anyone who carries a firearm is duty bound to train with it until the highest level of proficiency and safety has been achieved. After all, it is the tool of your trade.

Practice is the key to good shooting, but it must be the right type of practice if it is to be meaningful. Trackers must find the time to practice and excel at shooting from all possible firing positions and under all types of conditions. Anyone can hit a stationary target from a prone, supported position, but trackers have to hit a fleeting target visible for only a flash and often while on the move themselves.

When told that golf was a game of luck, Gary Player, the veteran South African golfer, replied, "Maybe so, but the more I practice, the luckier I get." This succinct response sums up the road to good shooting proficiency. Practice, and more practice.

Most shots taken by trackers during operations are from the standing position; therefore shooting from this position should be prevalent during training sessions. Next in line are shots taken from the kneeling position, and lastly from the prone position. Remember that in grassy, vegetated areas, a person lying down has a restricted field of view and is unlikely to see targets, particularly if they are lying down too. Therefore for shooting training to be mission oriented, shooting from the standing position should have priority.

During my military service in the South West Africa Territorial Force, I commanded a company of about 200 indigenous soldiers from the Caprivi Strip, which is situated on the Zambezi River between Zambia, Zimbabwe, and Botswana. One of our missions was to conduct interdiction operations along the border with Angola to prevent incursions of Communist-trained terrorists of the Peoples Liberation Army of Namibia from infiltrating the local population. One of the concepts that I desperately tried to instill in my unsophisticated troops was to "stand high, shoot low" so that when they made contact with the enemy, they had a better chance of success. Needless to say, when contact was made they all hit the deck and, sightless in the long grass, proceeded to shoot the leaves off the tops of the surrounding trees in typical Third World fashion. As Col. Jeff Cooper of the Gunsite Firearms Academy and dean of modern pistolcraft so aptly said: "The history of gunfighting fails to record a single fatality resulting from a loud noise."

Marksmanship Training— Retaining the Interest

Shooting practice can get boring if it is unplanned, so it behooves the team leader or training officer to devise creative and innovative shooting programs relevant to the task. Training must revolve around two operational facts: most contacts that take place between trackers and their quarry are usually at very short distances, and most are surprise targets. "Quick-kill," a technique devised by the military to engage short-range, opportunity targets is an excellent method and should be mastered by all trackers. (See Chapter 11, Training the Tracker, for an explanation of this and other shooting techniques.)

Shooting training for trackers can be made much more interesting by the choice of targets used. The constant use of the same old military or police targets can lead to serious identification problems later. British Field Marshal Bernard Montgomery used to say, "What you do in training you will do in battle," and anybody who has both trained and led troops into action knows full well the implications of this maxim.

This can be applied to shooting targets: if you continually use the same type during training, you will subconsciously expect to see something similar while on actual operations.

If anything appears to be different from what you expect, you will have a natural tendency to hesitate before firing. The criminal fugitive, recognizing your uniform, will have no such problem, and that gives him the edge over you. Any delay in firing may have fatal consequences to yourself and other members of your team.

During the Rhodesian war, many terrorists escaped because soldiers hesitated before shooting when they did not recognize them as the enemy. To overcome that problem, I used to get hold of as many terrorist uniforms as possible and use African soldiers to model them in front of incoming units so that field troops got to know exactly what to look for. Our kill rate escalated as a result. Correct target acquisition and recognition are essential prerequisites for operational success.

Probably the best way to overcome target staleness is to use a variety of types during training. Even the military silhouette target can be dressed up in old clothes to alter its appearance and make it more natural looking. If your budget can stretch to it, the best types of targets are the three-dimensional, polystyrene mannequin models, which are currently available at reasonable prices. These can be dressed up to add realism, and they are designed to fall down if struck in the head or vital areas of the torso. Even a near miss will not provoke a reaction—the shooter has to place his shots into one of the vital areas from whatever angle he happens to be from the target. If his shot is a good one, he is rewarded by the target's falling down in a realistic and satisfying manner.

A controversial aspect of live-fire engagements is when targets are hidden from view, which is usually the case when contact is made after a follow-up. Experience from the Viet Nam War clearly shows that the "spray-and-pray" technique so often used there proved the old maxim that "friendly fire isn't!" This dangerous technique has two predictable results. First, the enemy is rarely hit, and second, ammunition runs out rapidly at a cyclic rate of 600 rounds a minute. Because they tend to travel light, trackers do not have the luxury of endless supplies of ammunition, so they have to make their shots count the first time.

The "cover shoot" technique (illustrated on page 144) was devised to engage targets hidden from view in a systematic and economical way. With the aim of improving the shot-to-hit ratio, this useful technique is based on the fact that in a close-quarter firefight, 99 percent of combatants seek to hide from incoming fire by hitting the ground and rolling into the nearest cover. Accepting this fact, the cover shoot concept requires that two rounds be placed into positions of likely cover until all positions are neutralized.

Trackers, due to their basic team formation, can quickly move up into line and, using this technique, rapidly neutralize any firing coming from the cover to their front. By placing the shots low into the position, dirt and stones will spray up into the faces of anyone hiding there, causing them to take rapid evasive movements and thus exposing them to aimed fire. The trick is to try and place the bullets just above the ground, because a man lying down is no more than 12 inches high. To shoot any higher will result in the bullet winging harmlessly overhead. A four-man tracking team can quickly and effectively clear 40 potential firing positions, assuming that each man uses a 20-round magazine on a semiautomatic weapon. As a tracker called out to the scene of many contacts between security forces and terrorists, I was constantly amazed at the amount of bullet strikes high up in trees, some of them in excess of 10 feet. By training trackers in this technique, you not only increase your chances of hitting targets, there is also a dramatic reduction in ammo expenditure.

To prove the validity of this technique, line up your team and fire a shot over their heads. Ask them if they felt anything. Upon receiving a negative answer, ask them whether they mind if you repeat the exercise by firing a shot into the ground in front of them. I don't think you will get many takers!

During the Rhodesian war, one of our

tracking teams on the trail of three bad guys was hit by a sudden but heavy thundershower that quickly obliterated the tracks. Knowing that they were close to their quarry, the team continued in the same direction for several hundred yards. Moving carefully through the trees in the heavy rain, the trackers noticed movement ahead and observed the terrorists sheltering under a tree. Creeping as close as they could without giving the game away, the team opened fire. Three shots, three kills. Had the shots missed, the gang would have likely scattered into the forest and escaped in the rain. Just as important as tracking, therefore, is the ability to shoot accurately and effectively. Without it, the follow-up would be pointless. Again, the key is to create realistic, interesting training scenarios using a multitude of target types in a variety of terrain and vegetation settings.

TACTICAL AWARENESS

In his 2,000-year-old book *The Art of War*, Sun Tzu states, "If you know the enemy and you know yourself, you will not fear the outcome of a hundred battles." Why is it that as soldiers or policemen, we tend to forget this timeless wisdom time and time again? How many firefights have been lost because the participants failed to incorporate this concept into their combat preparations and plans? One only has to recall the recent debacle in Somalia when young American soldiers, all from elite special operations units, were unnecessarily slaughtered by an aggressive and competent foe fighting on his own turf, the alleyways and backstreets of Mogadishu. There is no doubt whatsoever that the troops fought with outstanding bravery and skill, as was indicated by the award of two Congressional Medals of Honor, but the questions must be asked: why were they there in the first place, why were they so lightly armed for such a risky mission, and where was the backup if required? In the same vein, what about the shootout in Waco, Texas, when brave, well-trained federal agents were ordered to enter a compound knowing they were outgunned and that their mission, predicated on surprise, was known in advance by the defenders? Both are examples of monumental intelligence failure, thumbsuck assessment, and catastrophic planning. All persons—military, police, or corrections—whose jobs may put them in harm's way *must* read and understand the writings of Sun Tzu if they wish to "not fear the outcome of a hundred battles."

Trackers should never lose sight of the principle, "Know your enemy, know yourself." Accordingly, they should make every effort to get to know and understand the limits of their own capabilities as well as those of the other members of their team. Once you accept those limits and know your capabilities, the time is right to develop the vital success factor of tactical awareness.

As you will have gathered by now, operational tracking is a team effort, and like any other team activity, practice and training together are the keys to success. By practicing basic team tactics and immediate-action drills, a team will start to work together like a well-honed, competent unit. Small-unit tactics are not difficult to master, certainly no more than basic football plays. Once the basics are mastered, more complicated drills can be attempted. Tactical awareness is developed by training to the point where actions and reactions require the minimum of thought and planning and become second nature.

Good tactical awareness is also achieved by the *desire* to soak up knowledge of tracking and tracking-related skills and operations. Talk to other trackers, discuss tactics, read books and manuals, watch training films, practice bushcraft, and stay alert to the operational situation in your district. Go over previous operations, seek out lessons learned, evaluate performance and tactics, learn from any mistakes, and discuss how you would do better next time. Remember, no man is a total expert—there is *always* more to learn.

An important route to tactical awareness is developing and retaining trust between you and your fellow trackers. Compare this with the trust and faith between high-wire trapeze artists. They most certainly would not perform with a person who has not earned their trust. After all, what would happen if a team member failed to perform as required? So it is with a tracking team. Constant attention to details and performance builds trust and confidence. Training, training, and more training with the same team members will create the right amount of trust so when faced with a difficult tactical situation, they instinctively know how each will react and perform. During my Rhodesian tracking days, for example, I had a tracker on my team (let's call him Frank) who was none too bright, but his tactical awareness was outstanding. It was almost uncanny how he would position himself, without orders, exactly where I needed him to be. A very rare but useful type to have around during operations.

Equally important to developing good tactical awareness is having a total understanding of your equipment, radios, weapons, and transport as well as your ability to operate them. Many operations have been doomed by personnel failing to understand exactly what equipment can and cannot do. Remember the ancient story of the knight who lost his horse when a shoe fell off because a nail fell out? Because the horse and rider were put out of action, the battle was lost, and so was the kingdom!

It is a prerequisite that your equipment be as good as your department can afford, be suitable for the task, and be tried and tested to your total satisfaction. I recall, with shame, a follow-up on which I was engaged in Africa, where equipment failure and my failure to test it played a significant part. My team was called out unexpectedly to follow the tracks of a group of guerrillas that had crossed the border of a neighboring country for sanctuary. At that time I was breaking in a new pair of Portuguese-made boots of which I was initially most impressed. Believing the boots to be up to the task, I decided to keep them on.

We were uplifted by helicopter and set out for the border, where we picked the tracks and commenced a hot pursuit. Within half a mile my feet were sore, within one mile they were raw, and within two miles they were a bloody mess, with blisters and bleeding flesh on the top, bottom, and sides. To add to my misery, the deep sand underfoot worked into the raw places, causing additional abrasion and discomfort. Eventually I cut up the boots and threw them away, continuing barefoot. Needless to say I was soon *hors de combat* and was ignominiously medevaced back to base. My team eventually made contact with the gang, wounding and capturing a senior guerrilla leader as well as recovering intelligence-rich documents and a large amount of cash! The point is, test your equipment, test your weapons, and test your communications until you are 100 percent satisfied that they will see you through thick and thin. Equipment breakdown is a poor excuse for operational failure.

The essential elements of trust, training, tactics, and testing go a long way toward attaining the elusive goal of tactical awareness, but as a tracker you still have to consider the skill and abilities of your quarry if you are to fulfill Sun Tzu's dictum. It is immaterial whether your team is police, corrections, or military—you must *always* know who and what you are tracking. You must attempt to obtain a full description of clothing, footwear, physical characteristics, weapons (if any), tactics, criminal history, and an assessment of the quarry's likely reactions should contact be made. With this information stored in your memory along with good training and tactics, good equipment, and a well-trained team, you *will* have the edge that Sun Tzu promised and "need not fear the outcome."

Tactical awareness will enable you to do the right thing in the right place at the right time, with the right tactics, to the right people, with the right effect, for the right reasons.

KNOWLEDGE OF LOCAL CONDITIONS

Part of being a good tracker is being able to fit into the terrain in which you operate. The United States contains more diverse terrain and vegetation types than any other country on earth, from the mountains of the Rockies to the swamps of Florida, from the deserts of Nevada to the tundra of Alaska, from the Badlands of the Dakotas to the bayous of Louisiana, from the rain forests of the Pacific northwest to the hardwood forests of the Appalachians, and from the prairies of the Midwest to the pine barrens of New Jersey. America has just about everything except jungles, and even then, troops who have been lost in the old growth forests of Fort Lewis at Tacoma, Washington, may dispute that!

Despite this fact, it is unlikely that a tracking team from California's Tuolumne County, situated in the foothills of the Sierra Nevada Mountains, will ever be expected to operate in the high desert country of, say, Utah. It is just as unlikely that a corrections team from Walla Walla Prison in the state of Washington will operate outside the state. The point is, every area has its own characteristics and peculiarities, and the local tracker must feel at home in that environment and be able to use it to his advantage. (This does not apply to military teams, as they could be expected to operate worldwide.)

Let me give you an example from the war in Rhodesia to illustrate this important principle. A game scout once spotted a large group of terrorists crossing the main road that runs through the wild, uninhabited Zambezi Valley, which separates Rhodesia from its northern neighbor, Zambia. From the terrorists' direction of travel it was obvious they were heading toward one of the few permanent waterholes situated along the foothills of the Zambezi Escarpment some 30 miles to the south. Because temperatures there permanently hover around 110 degrees, water would have to be a priority for the group. Knowing exactly where the waterholes were situated, the military authorities quickly but quietly placed ambushes along the game trails leading to specific watering points, and trackers were inserted by vehicle onto the gang's spoor. Realizing that they had been compromised, the now thirsty gang desperately blundered through the bush from waterhole to waterhole, losing several men to the ambushes. Severely affected by the lack of water and harassed by wild animals drawn by the scent of fresh blood, the group quickly lost all interest in continuing the struggle for liberation. They scattered, only to fall prey to aggressive tracking teams.

The lesson here is obvious. Trackers must have a complete and accurate knowledge of the area in which they expect to operate. In this case, the Rhodesian Army had already researched and plotted all permanent and seasonal water sources, so it was not difficult, based on the gang's direction of travel, to anticipate exactly where it could be intercepted. Trackers therefore should keep records of such things as water points, likely hiding places, isolated cabins, caves, mines, food sources, and areas where sympathetic human assistance may be given. Almost every sheriff's department has such records, but they must be updated continually to be of any real value.

Also important for trackers is the ability to understand and interpret animal and bird behavior. An alert tracker in the woods can pick up changes in activity and calls that differ from normal and may signify the presence of humans. I know of a wounded terrorist who was picked up because an alert tracker noticed a bunch of cows staring into a thicket. Investigation revealed him trying to hide in an ant bear hole. One for the cows!

Knowing local conditions can be directly related to a team's specific mission. Teams engaged on marijuana or moonshine search-and-eradication tasks, for example, should have a good knowledge of where these clandestine operations are carried out, the type of soil best suited for the cultivation of marijuana, and likely travel routes used by growers and distillers. Reports from all over the country indicate that marijuana growers are becoming increasingly more skillful in concealing their growing areas and irrigation systems, so it makes good sense to use trackers on eradication missions. This has the benefit of adding a pair of trained eyes to search for tracks leading to grow sites and to seek out any booby traps concealed there to catch the unwary.

Spencer Chapman, a British military veteran of jungle operations against the Japanese during World War II, wrote an excellent book of his activities entitled *The Jungle Is Neutral*. The jungle may be neutral, but it will certainly assist the tracker and give him a definite edge if he is able to manipulate the environment to his advantage.

CAMOUFLAGE AND CONCEALMENT SKILLS

One only has to look through any one of dozens of hunting magazines or sporting goods catalogs to see how important camouflage and concealment is to the hunter. Recently I counted more than 40 different camouflage patterns in only three catalogs, with a pattern for just about every vegetation zone and season, from swamps in summer to snow-covered forests in winter. Add to those all the different patterns worn by armies around the world and you will have an idea of the variety available.

Exactly the same amount of care and application should apply to the tracker too. In fact even *more* attention should be given to camouflage by the tracker to prevent his quarry from seeing him prematurely because, unlike a deer or antelope that would flee if it knows it is being followed, a human prey may have it in mind to attack his pursuer to effect his escape.

Obviously a tracking team on a follow-up cannot change uniforms each time the trail leads it into a different vegetational zone, but it can take some simple steps to make it as difficult as possible to be seen by the quarry. (See Chapter 11, Training the Tracker, for more on this fascinating subject.) The ability to conceal your appearance and intentions until you wish to reveal yourself is an important factor in achieving a successful follow-up, especially if the fugitives are armed and dangerous. It must be remembered that for a tracker to do his job, he has to move, and that movement is the biggest giveaway of all. Despite this, the simple steps outlined below will reduce considerably the chance of you being seen prematurely by your quarry.

A basic checklist that a tracker must consider in order to conceal his presence and intentions includes the following:

- *Shape*: to conceal recognizable body outlines such as head, shoulders, and arms.
- *Shine*: to reduce the chance of being seen by removing or covering light-reflecting surfaces such as a watch crystal, binocular lens, compass face, personal jewelry, or knife blade.
- *Shadow*: to reduce the chance of being seen by avoiding casting a recognizable shadow onto open ground and by remaining inside shadowed areas to reduce highlighting contrasting surfaces.
- *Surface*: to blend large surfaces into the background by the application of disruptive patterns and light-absorbing materials and coatings.
- *Silhouette*: to keep off skylines and away from contrasting backgrounds.
- *Movement*: to reduce or conceal any eye-catching movement by remaining in dead ground or shadows and by maneuvering tactically.
- *Noise*: to take precautions against making identifiable noises of human origin.
- *Tracks*: to prevent leaving a trail that could be followed by any opposition with tracking skills.
- *Ultraviolet and infrared signatures*: to take all necessary precautions to evade detection from the plethora of inexpensive night-vision and heat-sensing technology currently available. This will range from washing uniforms with a special soap (e.g., Sport-Wash) that is free of ultraviolet (UV) brighteners, washing untreated camouflage clothing with products such as U-V Killer and acquiring special garments that prevent the reflection of both heat and UV radiation.

In general, a tracker's appearance must blend into whatever background he finds himself in at that time, so a disrupted-pattern uniform of browns, greens, and grays is ideal for all but dry desert conditions. As long as the tracker is wearing an appropriately camouflaged uniform, there is no need to add local vegetation to it as is recommended in army fieldcraft manuals. Adding vegetation, if done properly, is a time-consuming exercise and may delay the follow-up when stops are made to change or freshen up foliage.

It is advisable, however, to make an effort to disrupt the outline of the head because this part of the human anatomy, particularly when wearing a hat, is very recognizable even at a distance. A few pieces of burlap or shreds of an old camouflage uniform sewn onto your hat will break up the shape admirably and is all that is required in most cases. Camouflage cream or face paint applied to the face, hands, and other exposed skin areas is essential to create a 100 percent camouflage effect.

It is equally important to prevent your weapon, watch face, and other pieces of equipment from reflecting light, even moonlight, and revealing your presence. Reflected light from a signal mirror or heliograph can be seen from more than 20 miles away, so it is essential to darken or hide any piece of gear that may act as a reflector. I have seen many photographs of soldiers, particularly in Viet Nam, wearing camouflage uniforms but negating any advantage achieved by shoving a pack of Marlboros, complete with shiny cellophane wrapping, into the band around their helmet. This is a dead giveaway and a perfect aiming mark for enemy snipers!

Other items that reflect light are rifle barrels, magazines, aluminum canteens, maps and map cases, compasses, spectacles, sunglasses, notebooks, and a whole host of personal items that you only have to look around to identify. Cover them up or place them inside pockets or pouches until required for use. Even sweat shining off a face, black or white, can be seen from a distance. I have blonde hair, and helicopter pilots could always identify me from more than 1,000 feet up in the air. There is an old military saying, "Never go where you can be expected." To telegraph your presence in advance will only result in a welcoming party.

STEALTH

A good tracker, like all woodsmen, moves gracefully through the woods like an antelope and not like an elephant on the rampage. (Actually, an elephant can move incredibly quietly if it wants to.) It is difficult to move silently when you are tired, so fitness is important. Talking, whistling, humming, charging weapons, radio chatter, and so forth can all be heard from long distances if conditions are right and wind is in the right direction.

Once, on a follow-up, one of our trackers, who by virtue of his large ears was known as Wingnut, heard a noise ahead and halted the team. He described the sound as mechanical, possibly a weapon being loaded, but he was not 100 percent sure. Taking the necessary precautions, we eventually came across an exhausted terrorist armed with a Soviet PPSh submachine gun (which has a cyclic rate of fire of 900 rounds per minute!) who surrendered, fortunately without firing a shot at us. The noise Wingnut had heard was the weapon being unloaded by the terrorist, and the distance was almost 1,000 yards! Remember that in the woods, where there is none of the background noise found in urban areas, one's hearing is intensified. Incidentally, at night, sounds seem four times closer than they really are (or, put another way, seem four times louder).

Trackers use visual contact and silent signals to communicate with each other while on a follow-up (see Chapter 5, Team Tracking and Tactics). It is essential that the team practice these signals until all its members fully understand the entire language, at least enough to communicate all the situations and reactions that they are likely to be confronted with on a follow-up. The importance of accurate communications can be seen in the following story. While following a gang of nine terrorists, the tracks led to a remote ranch house in a small grove of trees. I sent a tracker, a new man not yet used to working with my team, to scope out the house. A few minutes later he returned and from a distance called out, "In the house." We all took this to mean that the terrorists had entered the house, so I called up the commander of a platoon of infantry support troops and informed him of the situation. Angered by the fact that one of his fellow officers had been wounded by the gang several hours earlier, he decided to assault the house using rifle grenades fired through the windows.

Waiting until I had moved forward to cover the main door, the infantry commander was just about to give the order to open fire when the agitated tracker who had done the recce ran up and asked what we were doing. After quickly briefing him on the plan, he paled and said, "No, there is an old couple in the house," which was what he was trying to communicate to me by his earlier message. Fortunately, all's well that ends well, but Mr. and Mrs. Rancher never knew how close they had come to witnessing, firsthand, an attack from the wrong side.

We quickly resolved our communication problems and never made the same mistake again. The tracker later became a valuable member of our unit. (The nine terrorists had, in fact, passed close to the house and deliberately walked through the cattle pens in an attempt to conceal their tracks, even to the point of dragging cattle along with them. They were all eventually accounted for in a series of minor follow-ups and ensuing skirmishes.)

As far as trackers are concerned, movement is probably the most likely way he will be spotted by his quarry. To close the time/distance gap, trackers often have to move faster than would be prudent under other circumstances. There is no real answer to this problem; the main mission of a tracking team is to close with and make contact with the fugitives as quickly as possible. However, enemy observation of the team's movement can be reduced drastically by the skillful use of route selection and cover, particularly shadows. Shadows can hide a multitude of movement errors, and it is wise to use available shadows as much as possible. It is accepted that the tracker will have to be right on top of the spoor wherever it leads, but a little care and attention to basic camouflage principles, noise discipline, and tactical movement will go a long way toward minimizing the chance of being seen prematurely.

When called out for a tracking operation, it not wise to smell like the perfume counter at Sears. Aftershave lotion, deodorants, colognes, and perfumed soaps can be smelled for some distance, particularly when the wind is blowing in the right direction. A person who has spent several days in the woods quickly acquires a good sense of smell, and anything foreign to the environment is easily identified. Cooking oil, rifle oil, and burning stove fuel, particularly military-issue hexamine tablets, can reveal your presence long before you can be seen. So leave the Old Spice, Camay soap, and Big Macs at home where they belong.

A small patrol on its way to an observation post under cover of darkness one night caught a whiff of human sweat mingled with rifle oil. Knowing that there were no other permanent residents in the area, the patrol carefully moved upwind until it established the source of the odors. Voila! Seven guerrillas, fast asleep, "protected" by a standing but dozing sentry. Score 7 to 0 for our side.

A good bow hunter, well-versed in fieldcraft skills of camouflage, silent movement, and smell and noise prevention has a better chance of bagging a big buck than a poorly skilled rifleman with a scoped .30-06. It is the same in man tracking. Use every advantage you can find, and manipulate your surroundings to suit your purpose. Take control!

SUMMARY

By now you will have discovered that there is no magic to the basic qualities and skills required of a tracker. There are no arcane skills handed down from father to son or the requirement of an apprenticeship from youth. It requires only a set of basic human abilities, possessed by almost everyone to some degree, that need to be refined, honed, and sharpened.

Tracking skills are inbred in us from our primitive past. All we have to do is retrieve the information and put it to use in the service of our communities, whether it is in the apprehension of criminals, the arrest of fugitives, or the recovery of missing children.

Chapter 3
TRACKING TECHNIQUES

A person skilled in tracking can consider himself as having reached the pinnacle of bushcraft.

Lt. Col. Ron Reid-Daly,
Commanding Officer,
Rhodesian Selous Scouts

In terms of sensory input, tracking is probably unmatched by any other human activity. All of the human senses are employed: sight, hearing, touch, and smell, as well as the less understood phenomena of intuition and "sixth sense." A mass of information must be gathered and analyzed, assumptions made, conclusions drawn, and a course of action decisively acted upon. Weather, wind, time, direction, ground conditions, temperature, terrain, number of tracks, physical descriptions, condition of fugitives, whether armed or unarmed, and obstacles are all constantly fluctuating factors, any one of which could influence the success of the follow-up. Add to this mix the elements of danger, stress, tension, and fatigue as well as the need to continually assess and update likely courses of action taken by the fugitives and your own countermeasures.

Finally, resupply, water, food, ammunition, and communications must all be taken into consideration to complete the welter of human activity that characterizes a follow-up. To put it another way, a follow-up is an entire military operation in microcosm.

Following a man-made trail is like solving any other complicated problem. The solution lies in systematically identifying all individual factors, subjecting them to analysis, recognizing them for what they are, and fitting all the pieces together. By starting at the beginning, proceeding systematically and logically, and working it through to the end, a successful apprehension can be made.

WHAT TO LOOK FOR—
INDICATORS AND TYPES OF SPOOR

Regardless of ground type or terrain, there are three types of indicators that prove the passage of a human being or animal across a piece of ground. These are ground spoor, aerial spoor, and sign. Each has entirely different characteristics. When you are following the tracks left by one or more people over natural ground, you will always find a combination of these indicators, sometimes more, sometimes less of one than the others, depending on the ground surface and vegetation conditions.

Ground Spoor

Ground spoor is the term used to describe the marks, imprints, or indentations left on the ground that can be positively identified as being made by the person or persons you are tracking. It can be said that spoor is conclusive evidence of the people you are tracking, and ground spoor is the easiest and best type of spoor to follow. Examples of different ground spoor types include the following.

Flattening

Flattening is the pressing down or leveling of soil, sand, stones, twigs, or leaves by the weight of the body transferred to the ground by the foot, leaving an impression. Flattening is more likely to be found in hard, dry, sandy conditions where there is no moisture in the soil to hold a clear and lasting imprint. It shows up as a lighter color due to the difference in the way the light is reflected. Some examples of flattening are footprints in soft sand, an imprint of a knee or buttocks in yielding ground, the imprint of equipment such as a heavy pack or rifle butt, and the impression left in grass or sand by a sitting or lying person.

In the animal kingdom, very few creatures have flat, wide, long feet like a human's. Most large animals in North America, with the

Close examination of the ground behind the log reveals a fragment of moss scraped off by a boot plus a partial heel imprint left in the soil as all the weight of the fugitive is concentrated in the one spot.

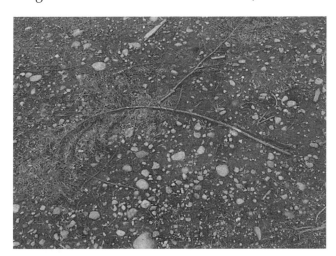

The twig in this picture has been pressed into the ground by the weight of a human foot leaving an imprint in the soft, damp soil. Subtle clues like this are often the only ones available to a tracker in coniferous forests.

On mossy, yielding ground, a tracker should look for twigs either broken or kicked out of place. Animals with their small, sharp feet will not leave sign like this.

exception of bears and felines, have cloven hooves that make characteristic prints on the ground.

Regularity

Regularity is the eye-catching effect of unnatural, man-made patterns. Man-made footwear has several characteristics that differ from the bare sole. One of these is a raised or cut pattern designed to enhance traction on slippery or wet surfaces and is found on virtually all synthetic soles. It is these patterns that make tracking so much easier, as they leave distinct marks on the ground that are not present in nature. The eye tends to pick up these patterns because they are unnatural to the environment.

There are dozens of different designs, lines, circles, zigzags, whorls, or other geometric patterns depending on the nature and purpose of the type of shoe involved. The sharp outline of a plain sole or heel also shows up clearly, as does a bare footprint with its rounded heel, arch, and toe marks.

An example of regularity is a boot imprint left on soft, muddy ground. The sharper and deeper the tread pattern of the boot, the softer the ground, and the heavier the person, the more obvious the spoor will be. The damper the soil, the clearer the imprint.

Print regularity can be left when a fugitive exits a muddy stream or puddle and remnants of mud corresponding to the print outline are deposited on the ground in the same way as a lino cut on paper. A person coming off wet ground into a dry area will leave prints, but these will eventually dry up and disappear. Prints can also be left on dry, dusty surfaces, particularly along dirt roads and in abandoned buildings.

The more detailed the pattern the easier it is to see. The heavily cleated Vibram sole catches the eye immediately, but the military Panama pattern to the left is not so obvious. A plain, unpatterned sole would be even more difficult to see.

Aerial Spoor

Aerial spoor is the path seen by the displacement or damage done when the person being tracked enters an area of vegetation, be it grass, scrub, or bushes. The direction of travel can be clearly identified by the direction and angle in which the grass is bent.

Aerial spoor is not confined to the ground but can often be observed on trees and bushes up to head height. It is not always conclusive evidence in the same way as ground spoor because it cannot always be confirmed as having been made by the fugitive. A better description is that it is substantiating evidence—it is insufficient in itself, without other confirming factors, to prove that it was made by the person or persons you are tracking. As a trainee tracker this is important to remember, because it is very easy to "seize" on an isolated aerial spoor in the belief that it is part of the real trail and wander off in the wrong direction.

Aerial spoor, while inconclusive in itself, can often provide good clues about the passage of clumsy human feet. The broken stick and the dislodged leafy twig will catch the tracker's eye, and a closer examination of the ground beneath will often provide the confirming evidence—in this case a toe print in the loamy soil.

When a human passes through thick vegetation, plants will be dragged in the direction of travel. Often the lighter colored underside of the leaves will contrast against the background and attract attention.

Some examples of aerial spoor are dragged vegetation displaying the paler underside of the leaves, snagged vines pulled in the direction of travel, displaced bark revealing a lighter color, crushed and bent grass, broken and bent vegetation, stripped leaves, crushed twigs, overturned leaves, skinned bark, knocked-off dewdrops, disturbed insect nests, and broken spiders webs. When following aerial spoor, the tracker must constantly look for ground spoor or other forms of confirmable evidence to assure himself he is still on the right tracks.

Sign

Sign refers to a whole host of other indicators that do not fall into the categories of ground or aerial spoor. Examples of sign include stones that have been knocked out of their original position, overturned leaves showing a darker underside, sand deposited on rocks, drag and scuff marks, displaced twigs, and scuff marks on trees.

It is critical that a tracker does not use sign in isolation; it can be made by a multitude of animals or birds and not necessarily by the quarry. When on a follow-up, the tracker should constantly look for confirmable evidence to assure himself he is on the correct set of tracks. Still, there are times when, due to the nature of the ground, this is not possible. The tracker then must use a combination of instinct, assumptions, and inconclusive clues. If the quarry has been constant in his direction of movement and the tracks disappear, it is more than likely that he will continue along the same line. This is when the tracker has to rely on sign, which, while not the best, is all he has. Very often this is enough to go on until a piece of confirmable evidence is found.

Litter

Items such as paper, packets, cigarette butts, and discarded bandages are known as litter. They are neither spoor or sign, but unless found on the track itself, they may or may not be related to the follow-up. Paper or other litter can be blown by the wind and may not necessarily be discarded or dropped by the fugitives. If litter is found that is suspected of being related to the follow-up, other evidence should be sought before a conclusion is made.

It is important to remember that no individual trace of spoor or sign ever occurs in isolation. Every spoor or sign has a before and an after, meaning that to leave an indicator, a person has to get there and has to leave there. This may appear to be overstating the obvious, but many times on training courses students have found a mark of some sort and convinced themselves that it is of some significance to the follow-up. This usually happens when they have become bored while searching for a lost spoor and their imagination takes over. In a tracking situation, focused mental power is an asset, but a wild imagination can be a definite drawback. If an isolated indicator is found, it is imperative that a close examination of the entire area is undertaken to either confirm or reject the discovery as relevant to the follow-up. When on a follow-up, a tracker will, in most cases, come across a mix of indicators unless he is lucky enough to follow tracks along a beach or after a snowfall. The typical follow-up will pass over a wide variety of ground and vegetation types as well as through different light and shadow effects, so the tracker will have to keeps his wits about him and focus his complete attention on the trail.

THE RULES OF TRACKING

Tracking, like any other human activity, has a set of basic rules. These rules are not made arbitrarily by some nebulous "International Brotherhood of Tactical Trackers" but are based on long experience and trial and error. If you intend to become a tracker, you should read, learn, and digest these rules for yourself because failure to follow them will lead to failure on the follow-up.

Rule 1. Identify the correct tracks you wish to follow.

This might appear to be overstating the obvious, but on several occasions that I am aware of, trackers set out in pursuit only to find that they were following tracks of people who had absolutely nothing to do with the follow-up. Even if you have to spend additional time to ascertain the correct tracks before commencing a follow-up, it is worth it.

Rule 2. Mark the start point of the follow-up.

This is known as the "initial commencement point" of the follow-up. It may be at the scene of the crime, at a point where tracks have been intercepted, or where fugitives were seen by a witness. This is important because if the spoor is lost, you may have to return to the commencement point and either start over or attempt to backtrack.

Rule 3. Never walk on top of the spoor.

Stay off the tracks by walking to one side. There are several reasons why you may have to return to a certain point along the trail, as Rule

4 will indicate. So do your best to conserve the spoor at all times.

Rule 4. Never overshoot the last known spoor.

This is the most important rule in tracking but the one which is most frequently broken. The last known spoor is the last indicator that the tracker can see ahead of him. It is vital that he stay behind the last known spoor so that he can always see the direction in which the tracks are heading to keep the momentum of the follow-up going. Should he overshoot the last known spoor he may fail to see any more indicators ahead and will have to stop the follow-up while he conducts lost spoor procedures.

Rule 5. When following aerial spoor or sign, always check for confirmable evidence.

It is easy to wander off the correct trail if a tracker is compelled to rely only on substantiating evidence such as aerial spoor or sign. To assure himself that he is on the correct tracks, he must continually look for such confirmable evidence as ground spoor or other positive indicators.

Rule 6. Always know exactly where you are.

When following armed and dangerous fugitives, it is essential to know *exactly* where you are should you wish to call in other teams, arrange a medical evacuation, or call for a resupply of food, water, or ammunition. Valuable time will be lost if a pilot or vehicle driver has to search for a tracking team that has sent an incorrect location.

Rule 7. Always keep visual contact with other team members.

The integrity of the team is maintained by constant visual contact between the members. In the event of an armed confrontation or surprise ambush, the controller must know exactly where all team members are located so that he can exploit the tactical situation by employing the correct immediate action drill and ensure the safety of the other team members in the event of a firefight. An "own casualty" is a very difficult thing to live with and even harder to explain to a grieving wife and family.

Rule 8. Always try to anticipate what your quarry will do.

By carefully observing the actions of the quarry based on the evidence of the spoor, it is possible to "get into" his mind and anticipate what he is likely to do under certain conditions. This enables steady and continuing progress during the follow-up.

FOLLOWING THE SPOOR

If you go for a walk in the woods and strike out in a straight line, you will notice that your path will cross over a spectrum of ground types, including open bare patches, low and tall grass, areas of rock and stony ground, marshy meadows, bushes, groves of trees, streambeds with moss-laden rocks, and a whole host of other conditions. Some areas may be in permanent shadow under tree canopy, open patches may be exposed to constant sunlight, wind, and rain, and there is always the possibility of crossing man-made or animal trails.

In places along the trail where the ground is soft or moist, you will find that you leave footprints; in other places you will find that your passage will be characterized by crushed and broken vegetation. All of this will be influenced by the type of ground you are on, sun angle, dark or light weather conditions, and other factors. Whatever the circumstances, the point is that the tracker has to cope with a wide range of variables. His attention must be sharply focused if he is to maintain momentum and close the gap with his quarry.

How to See the Spoor

Most people, when they think of tracking, believe that they must locate and identify every single mark and sign as they progress along the trail. This is not so. To track effectively and in such a way as to cut down the time/distance gap, operational trackers must get in the habit of looking ahead, often up to 30 to 40 feet or so, and seek the trail there (Figure 1). Compare this to riding a bicycle. If you ride along looking at the road close to the front wheel, you will make little progress, but if you lift your eyes and scan ahead you will be able to see both the obstacles and clear routes and speed up considerably. There are times, however, when the going will be hard and you will have to track close to your feet, even getting down on your hands and knees in search of minute evidence of spoor or sign.

Tracking continually close to your feet has five undesirable effects:

- It is time-consuming and slows down the follow-up.
- The tracker on the spoor loses the "feel" of the surrounding country.
- The tracker loses visual communication with his flank trackers.
- The tracker becomes prey to possible ambush.
- The tracker tires quickly from eye strain.

By scanning ahead, the tracker is able to continually cover the ground and be aware of the behavior and movement patterns of his quarry. He also remains in visual contact with his flank trackers, increasing his safety. Also, by tracking ahead, the tracker does not have to waste time searching for close-in spoor. He seeks for upcoming spoor and quickly moves up to that point, all the time looking ahead for the next indicator. This way, he tends to leapfrog along. Under certain ideal conditions when the ground is free of vegetation and the spoor clearly visible, it is possible to run on the tracks, which, if safe to do, dramatically reduces the time/distance gap.

So trackers should get into the habit of looking ahead. It's amazing what they can see when they raise their eyes and look up. Trainees on a basic tracking course always find it hard to look ahead and tend to concentrate on searching for tracks close to their feet. It is always gratifying when they finally break this bad habit and suddenly "see the light."

Using the Light to Your Advantage

The main reason we are able to see tracks is because light from the sun casts a shadow in the imprint of the sole pattern on the ground. It follows then that footwear with a deeply cut sole pattern will cause deep imprints on soft ground and therefore cast more distinct shadows, and shallow sole patterns will create less distinct shadows.

It also follows that the lower the sun angle

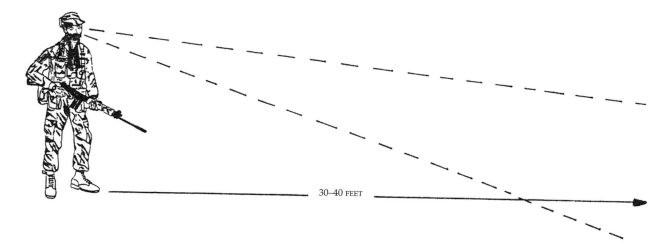

Figure 1: Seeking the spoor. The tracker should not attempt to search for the spoor directly at his feet. He should look ahead to the farthest point where he can still see the tracks. This enables him to move quickly and retain his "feel" for the environment.

relative to the footprint, the more distinct and obvious the shadow will be. As the sun rises to its peak, the shadow in the print will become less distinct until at midday, with the sun directly overhead, it disappears completely. With the passing of the day, the sun, moving down to the horizon in the west, will cause the shadow effect to reappear, making the print visible again (Figure 2).

With this in mind, the optimal time for tracking is from early to midmorning and midafternoon to early evening. The worst time is from shortly before to shortly after noon. As combat trackers we must never let this influence the follow-up, which must go on nonstop irrespective of the time of day, until contact with the quarry is made. Tracking is a little more difficult during the midday period, but if the tracker sticks to the basic principles progress will be made. Bear in mind that your quarry may want to stop and rest during the heat of midday, giving the

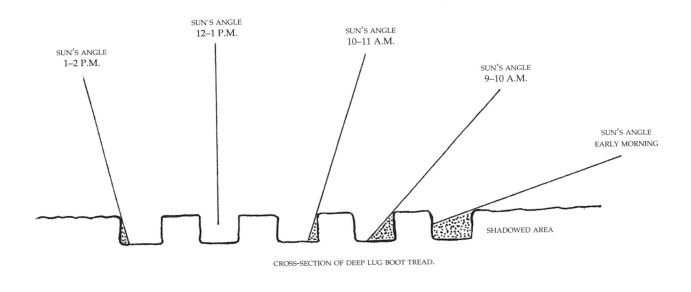

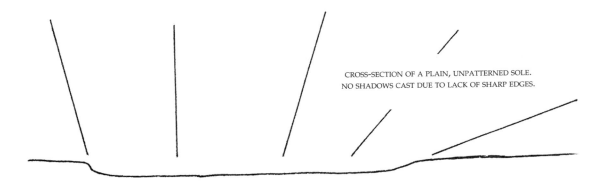

Figure 2: The time shadow effect. Note that the shadow effect is more prominent in the early morning when the sun's angle is low. As the sun moves higher in the sky toward the midday, the shadow diminishes until it can no longer be seen. As the sun drops toward the horizon throughout the afternoon, the shadow reappears. Therefore, the optimal times for tracking are early to midmorning and late afternoon.

tracker the opportunity to cut down the time/distance gap.

The tracker must learn to take advantage of the sun's angle and look for the spoor from a position where the prints are between him and the sun. This way the shadow effect is maximized and the prints are easier to see. To illustrate this, place a line of prints on clear, soft ground and walk around them in a circle. You will see that at certain angles the shadow effect is greater and the prints seem to stand out. If you look up you will see that the prints will be directly between you and the sun. By tracking from this angle, even if you have to look back over your shoulder, your progress will be better and faster.

As has already been stressed, operational trackers must keep the follow-up momentum going in an aggressive way whatever the light conditions. Very often there will be no sunlight at all, or the trail will pass through shaded areas, making the tracks difficult to see. Tracking is still possible in these conditions, and with practice you will overcome these temporary handicaps.

If pursuing armed and dangerous fugitives, under no circumstances should tactical trackers operate at night. Search-and-rescue tracking, however, can be carried out in the dark using flashlights, lanterns, or car headlights as a substitute for the sun. Master tracker Jack Kearney of the U.S. Border Patrol describes how to use this technique on page 115 of his book *Tracking: A Blueprint for Learning How*.

The Tracking Stick

A useful tool used by nontactical and search-and-rescue trackers is the tracking stick, which was developed by Jack Kearney. The tracking stick, normally about four feet long and tapering in width from half an inch to a quarter-inch, is used to measure the distance between footprints on the ground. Let's say that the quarry being followed has an average pace of 30 inches, measured between the toe of the left boot and the toe of the right boot. This distance is marked on the tracking stick with an elastic band, string, or indelible marker. When the tracker cannot see the next footprint on the ground, he crouches down, places the mark on his stick in the center and along the axis of the last footprint, and examines the ground under the end of the stick. It stands to reason that the next foot to strike the ground will be very close to this point, and it is more than likely some evidence will be found there. If, after close examination, no mark is found, the tip of the stick should be rotated in an arc of approximately six inches to either side and the ground closely searched for evidence there. If no evidence is found, the arc is increased up to 12 inches to either side and the ground searched again. Some evidence of the next print should be found within the arc, but if it is not, the tracker should waste no more time and go immediately into his standard lost spoor procedures.

It must be stressed that the tracking stick has no place in tactical tracking because when following a dangerous fugitive, the tracker will be carrying a weapon in his hand. If, however, he finds advantage in using the tracking stick technique, the rifle should be used in place of the stick.

Put Yourself in His Shoes

A good tracker will always try to put himself in the mind of his quarry. He asks himself, "What would I do and where would I go under the same circumstances?" In this way he can quickly pick up the patterns and habits of his quarry. Once the pattern is established in his mind, he can anticipate very accurately where the next spoor is likely to be. For example, if the quarry keeps to high ground, prefers to keep under ridge lines, or likes to slink along streambeds, this will soon become known to the tracker. Similarly, if the quarry suddenly changes his tactics, the tracker will become alerted and start to pay special attention to find out why. By getting into the mind of his quarry, a good tracker can often anticipate exactly where the spoor is going and, by consulting his map, can maneuver spoor cutting teams, stop groups, or ambushes in front of the fleeing fugitives. (See Chapter 8, Command and Control of a Tracking Team.)

INTERPRETING THE SPOOR

By examining the way a line of footprints are left on the ground, we can gain an amazing amount of information. We can tell whether the person responsible for the prints is young or old, big or small, male or (sometimes) female, fresh or tired, in good health or bad, nervous or confident, moving slowly or quickly, and if he is lightly or heavily laden. Quite remarkable for someone we may not have met! This is done by examining and analyzing the way the foot impacted the ground, the amount of time it dwelled there, and the way it left.

The Dynamics of a Footprint

A normal-sized person, say 170 lbs., in good health, going from point A to point B with a purpose in mind, will leave a fairly regular line of prints. Assuming the ground is level, soft, and moist, each print will be about the same size and the steps will be the same distance apart. If we examine one of the prints closely, we will notice from the impression that the back of the heel has struck the ground first. The person's weight then revolves evenly through the heel to the sole, and through the sole to the toe, when the shoe leaves in preparation for the other foot to strike the ground. If we take this information as a baseline, anything that differs from it can be compared and analyzed, and a fairly accurate profile of the person will emerge.

When attempting to interpret clues left by a footprint, there are two important aspects to consider:

- *The primary impact point* (PIP): The point where the shoe or boot strikes the ground first and reveals the most information.
- *The terminal point* (TP): The point where the foot leaves the ground in preparation for the next step.

A careful examination of these points can reveal much information about what the person was doing at the time the prints were laid. For example, if the PIP of a line of prints is deep and the prints widely spaced, we can deduce that the person was moving quickly because the faster one walks, the deeper the heel mark, or the PIP, will be. Similarly, if the person slows to a stroll, the PIP will be less defined because more of the foot will strike the ground, spreading the impact over a wider area.

This phenomenon is based on Newton's Law of motion: a force is opposed by an equal and opposite force. What this means is that an object in motion carries kinetic energy, and the faster it moves the more energy is generated. Conversely, the slower an object travels the less kinetic energy it carries. In tracking terms it means that a person moving at a fast pace builds up kinetic energy, which is transmitted to the ground through the heel as it hits. (If you wish to experience this effect for yourself, make a fist and, from a distance of two feet, strike the wall slowly. The slow-moving fist carries little energy, so you are not hurt. Now try the same exercise by striking the wall as fast as you can. You will feel a marked difference!) Examine these effects for yourself in a patch of soft sand and you will understand the principle.

Assume that we have come upon an unknown set of prints and want to know something about the person who laid them. If we measure the length and width of the prints, we can assess the shoe size fairly accurately, and by looking at the print itself we can identify the type of shoe, its use, and possibly even the manufacturer. If we examine the way the foot impacts and leaves the ground, we can ascertain the speed at which the person was traveling when the prints were laid. If we measure the length of the pace related to the size of the shoe, we can tell whether the person was tired or sprightly. We can tell from any abnormal wear on the tread if he suffers from an ailment or injury. By comparing the depth of print with our own, we can assess whether the person is of large stature or carrying a load. We may be able to see from the spoor whether the person stopped and rested or whether he relieved himself, which always provides a number of good clues (for example if the quarry is male or female). Excreta, vomit, or blood spoor may indicate his state of health. If we follow the tracks to see where they go, we may even gain an idea of what the person was doing, and if we backtrack we can find out where he came from.

If we are lucky in closing the time and distance gaps we may actually come across our quarry. Even if we don't, by comparing and evaluating what we have observed we can come up with a fairly accurate picture of our mystery person and his activities. A tracker must develop the mind of a detective and search for all the clues left along the trail. The more observant he is, the more clues he will recover and the better he will be at assessing the state of mind and physical condition of his quarry. By doing so, it will provide him with the insight necessary to predict what his quarry may do and which appropriate actions to take.

Up and Down Slopes
- A man going up a steep slope will place the toe of his boot down first, and the heel may not be seen at all. The steeper the slope the more he will try to dig his toe in. If the slope is very steep, he may try to climb up at an angle across the face. If this is the case, the outside edge of one foot and the inside edge of the other foot will be apparent. Grass and other plants will be crushed or pushed up the slope. The steeper the slope the closer his steps will be.
- A man going down a steep slope will place his heel down first, which will be plain to see, but the toe may not be seen at all. Grass and other plants will be crushed down the slope, and there may be some evidence of sliding and skid marks. A person running downhill will leave widely spaced prints, but a heavily laden person tends to walk with a smaller, careful pace.
- A man carrying a pack down a slope may leave skid marks of his butt and disturb more vegetation, which will point downward in the direction of travel.

Physical Disabilities
- Tired or physically impaired people will walk with a shorter pace and tend to drag their feet, leaving clear marks.
- An injury to a leg will be indicated by the good leg print being deeper and the injured leg print less distinct or twisted out of normal alignment.
- A seriously wounded person may have to proceed on his hands and knees or be compelled to take frequent rests.

Carrying Weight
- Additional weight is indicated by closely placed, deep prints, and possibly by frequent rests in shaded areas.
- A person carrying weight will take the shortest and most direct route.

Speed of Movement
- A fast-moving person will leave widely spaced, deep imprints with distinct heel marks and a scrape mark on the ground made by the toe as the foot leaves the ground.
- Prints left by slow, deliberate movement will be evenly spaced and uniformly deep, with little difference between the primary and terminal impact points.
- The PIP of a person sprinting will be from the ball of the foot, with little to no heel marks left at all, and the prints will be far apart.
- Prints of a person creeping along will be either tippy-toe or closely spaced and even.

Disguising the Tracks
- If a person walks backward in an attempt to confuse a tracker, it will be obvious because the impact points will be reversed. The toe will impact first, pushing the soil back. When the heel leaves the ground it will scrape and drag sand or soil in the direction of travel. The prints of a person walking backward will be closer together than normal and often appear to be unbalanced.

Age and Sex
- Generally speaking, it is possible to tell the

age and sex of persons from their ground spoor. A child's prints will be obvious from the size of the footwear, and a woman's prints will usually be smaller and narrower than a man's. The type of shoe worn may also give an indication of the sex of the wearer. For example, high heels or fashion styles will likely indicate a woman, but carefully weigh all the factors before coming to a conclusion about sex. There are a lot of women who wear footwear common to both sexes, particularly when engaged in outdoor and sporting activities. It should also be noted that the United States does have some of the largest women in the world, and some men are of smaller stature, particularly those of Asian descent.

ASSESSING THE AGE OF SPOOR

Ground spoor, aerial spoor, and sign are all affected by time, wind, rain, sun, and frost in one way or another, and it is important that the tracker takes time to study these changes. Knowing how and why these changes occur gives the tracker a fairly accurate indication of the age of tracks. Note that when making an assessment of time, it is important not to use one single indicator in isolation, because this may lead to false assumptions. Far better to use a variety of indicators, which should give a more reliable estimation.

The Time and Distance Gap

As you are well aware by now, the aim of tactical tracking is to steadily reduce the time and distance gaps between the fugitives and trackers until the fugitives are apprehended or accounted for. The time gap is ascertained by assessing how long it would take for the trackers to catch up with the fugitives if the fugitives remained stationary. The distance gap is the actual distance, measured in miles, between the trackers and fugitives at a given point in time. For example, if the fugitives are stationary six miles ahead of the trackers and it will take the trackers one hour to cover two miles, then the time/distance gap is three hours. Similarly, if it takes the trackers two hours to cover one mile, the time/distance gap is 12 hours. The differences are because tracking speeds can vary considerably due to factors like time of day, weather conditions, physical condition of the follow-up team, countertracking done by the fugitives, difficult ground, and a whole host of other reasons.

Knowing the time/distance gap is important to a tracking team for several reasons. If a team is following up on a gang of armed fugitives, it will need to have an idea of when it is likely to catch up so it can plan appropriate tactics, for example. Similarly, if the fugitives are well ahead of the team, the command and control element may plan to use spoor cutting teams, aerial observers, or other resources.

As a follow-up commences, the team on the spoor will periodically inform its control by radio what it assesses the gap to be. The main reason for this is to see whether the team is getting any closer to the fugitives. A team that finds that the gap is increasing rather than decreasing must seriously consider other alternatives, such as leapfrogging (See Chapter 6, Alternative Tracking Techniques and Follow-Up Methods) or whether it should be withdrawn or replaced by a fresh team of trackers.

The age of spoor can change suddenly and dramatically. Take, for example, a situation where spoor is about four hours old. The fugitives decide to take a rest for three and a half hours, unaware that trackers are on their trail. When the trackers arrive at the resting place, the age has suddenly been reduced to 30 minutes. It is for this reason that the team should always track believing that contact is imminent so that a high degree of alertness is maintained.

Although great care must be taken to assess the time/distance gap, trackers can be way off on their assessment. On one occasion in which I was involved, it was last light and the follow-up had to stop due to coming darkness. On our final SITREP (situation report) for the day we assessed the tracks of eight guerrillas as being 24 hours old. Commencing the follow-up next morning, we hadn't gone more than a quarter-mile when we came upon an overnight camp on a riverbank showing evidence of a hasty evacuation, with open packs and foodstuffs lying around. In this case, the gang, not knowing that it was being tracked, had taken a day off to rest, allowing us to close the gap.

It is not necessary to be explicit in assessing the age of spoor. It is acceptable to work in two-hour increments. As a general rule, it is better to overstate the time/distance gap than understate unless you are fortunate enough to have an eyewitness who can pinpoint the exact time. However, if you feel that due to the state of the indicators a contact with your quarry is imminent, then you can talk in terms of minutes rather than hours.

Assessing the age of tracks is a skill that can only be developed by long experience, but there are certain factors that, if taken into consideration, can assist neophyte trackers in making reasonably good assumptions.

The Effects of Time and Weather on Ground Spoor

Sharply cut sole patterns such as the Vibram or lug type commonly used by the military and hikers will leave a distinct print in soft, damp ground due to the depth of the tread. The deeper the tread and the softer the ground, the more noticeable the spoor will be. It stands to reason that a shallow pattern will leave less of an imprint and a flat, plain sole will leave only a faint image of the edge of the heel and sole. No matter the tread, prints are held together longer in damp soil than they are in dry sand, and the harder the ground the less likely it is to hold an imprint at all.

When a footprint is left on soft, moist ground, it will be clear cut with sharp edges and have a dark appearance, but the drying effects of wind and the sun will cause the edges to erode and appear to become lighter in color over time. This is because the moisture holding the print together dries out, and particles of sand and detritus fall into the spoor or get blown away by the wind, thereby eroding and rounding out the once sharp edges and softening the shadows (see Figure 3). The longer a spoor is affected by wind and sun the less distinct it becomes until it disappears altogether.

A good way to illustrate this is to build a "spoor pit" of sand or soil and lay a set of tracks along one side. Over the next few days repeat the process and compare the difference in the prints. You will begin to get an idea of how tracks appear according to their age. There are several factors that must be taken into consideration, however. Tracks laid in the cooler temperatures of the night may appear to be fresher than they really are the next morning. Tracks laid in shady areas will always look fresher than those laid in an exposed, sunny place. Nonetheless, by studying spoor of known age over different types of ground, you will quickly get the idea and become fairly proficient at it.

Don't get too cocky, however, because there are times when even experts are fooled. In 1977, game rangers in South West Africa, while on patrol in a remote area of the Namib Desert, came across clear vehicle tracks winding through an area of massive sand hills. Fearing poachers seeking the horns of the endangered desert rhinoceros, they followed the tracks for several miles in their vehicle, only to discover a German field gun complete with trailer of World War I vintage. Later research revealed that the gun had been towed into the desert in 1914 by the Kaiser's soldiers to avoid capture by the British Army Expeditionary Force. Despite the fact that 63 years had passed, the dry, rainless desert conditions had preserved both the gun and its tracks.

The lesson here: always expect the unexpected.

The Effects of Rain

Rain, of course, has a serious effect on ground spoor, depending on the amount of precipitation and ground conditions. Heavy rain on soft ground will quickly wash out all traces and can be a major cause of follow-up failure. There is an upside to heavy rain, however: it leaves the ground fresh and clean so that any spoor laid after the rain can be followed easily.

Should a rain shower occur either before, during, or after spoor has been laid, it can give you a good idea of the age of the tracks if the time of the start and end of the shower is known. Raindrops on top of the spoor indicate that the rain fell after the spoor was laid, so if you know the time the rain fell it will give you an idea of the age of the tracks. During a follow-up, trackers should record times of rain showers for this purpose.

Light rain, sprinkles, or mist, on the other hand, can help a tracker by temporarily consolidating the soil, which may harden ground spoor when the precipitation ceases, especially if followed by hot sun. Heavy dew can also be an advantage, especially if deposited on grass and brush—footprints can be seen clearly on short grass, and the passage of a human will knock the dew off bushes and long grass.

The Effects of Wind

Wind has a very damaging affect on spoor because the edges of the prints erode faster due to the drying action and the displacement of sand particles. The stronger the wind the faster the tracks will disappear. Leaves, sand, and other detritus are often blown into tracks, giving them the appearance of increased age. The combination of wind and hot sun speeds up the erosion process considerably in all but the most shaded and protected tracks.

The Effects of Human and Animal Activities

Regular and predictable activities of both humans and animals can help a tracker establish the age of tracks. Say, for example, it was known that a farmer herds his cows to the pasture every morning and back to the barn at the same time every evening. This establishes a time frame to work on. If fugitive spoor was laid over the cattle tracks on the way to the barn, it indicates that the quarry passed by during the night. If the tracks were laid over both incoming and outgoing cattle spoor, it indicates that the quarry passed by during the day.

Most species of wild animals and free-ranging cattle, as well as most birds, drink in the morning and evening, so if animal or bird tracks are superimposed on a fugitive's spoor at a waterhole or river, it can indicate when the spoor was laid, especially if other indicators are taken into consideration.

The Effects of Time and Weather on Aerial Spoor and Sign

As is the case with ground spoor, time and weather conditions affect aerial spoor and sign. Vegetation that has been crushed or broken will wilt and dry out over time. Obviously the drying effect will be accelerated in hot weather or open sun but will be slower in cooler temperatures or shady areas. An understanding of these effects will help the tracker assess the age of the trail fairly accurately, but remember that aerial spoor and sign are not conclusive evidence, so any assessment of age must take this into account.

Figure 3: The effects of age and weather on spoor. A fresh print on soft, moist ground will leave a clear impression with well-defined edges. The sun and wind will rapidly erode the sharp edges, and sand will be blown in until the print is no longer recognizable. The rate of degradation depends on how much the print is exposed to the elements.

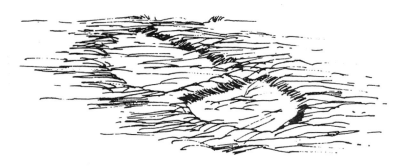

When grass, leaves, or branches are forced out of their natural position by the passage of a human body, nature has a marvelous way of restoring them to their natural alignment. By carefully observing and process of restoration, you will gain an understanding of how long it takes for displaced plants to resume their natural position. This is another way the age of a trail can be estimated.

Dry Grass

Passage of big, flat, human feet through dry grass will bend and crush the stalks down in the direction of travel. Obviously, the more people walking single file the greater the crushing effect will be. Whereas some greener stalks will eventually return to their original upright position, most dry grass stalks will stay broken and crushed on the ground until they decay or are burned off. Unless confirmed by other factors, dry grass is not the best method of assessing the age of spoor.

Green Grass

Green grass is a better indicator of age because it contains moisture. If moisture or dampness is present at the point of a break or where the stems are crushed and bruised, the tracks are very fresh. As time elapses, damaged stems will eventually die and dry out, with the drying process varying according to the ambient temperature, wind, and amount of direct sunlight.

Once, on an early morning follow-up, my team knew it was close to its quarry because the green plants and grass that had been crushed down onto underlying stones were still moist to the touch. Knowing the group we were following consisted of about 30 armed guerrillas, we alerted the local fireforce, which flew directly toward us along the back azimuth taken by the gang. This was done in an attempt to trap them in a typical "hammer and anvil" maneuver, with us being the anvil. In short order we heard the *wup wup wup* of the approaching helicopters and seconds later heard over the radio that the pilots had made visual contact with the scattering gang.

Spreading out as wide as a four-man team safely can, we advanced slowly toward the contact area ahead, where the choppers were pouring 20mm cannon shells into the position where the gang had gone to ground. Unable to make contact with the fireforce commander due to the noise and heavy radio traffic, we stopped advancing and hastily searched for some cover. Several seconds later the remnants of the gang burst through our "camouflaged line" hotly pursued by a barrage of cannon and machine gun fire. With shells exploding all around we hit the ground in terror lest the gunners mistake us for the guerrillas and open up on us. To my disappointment none of our tracking team, which had followed the gang for the previous 36 hours, was able to fire a single shot during the melee, but fortunately most of the gang fell prey to the airborne gunners and the fireforce troops. It was a surreal experience to find ourselves in the middle of a firefight and no one even knew we were there!

Leaves

When traveling through thick vegetation, men carrying packs, equipment, and weapons tend to tear off leaves and twigs from surrounding bushes, which fall to the ground. By comparing these fallen leaves for signs of freshness against those still on the plant, it is possible to assess the time they were displaced.

Dry, dead leaves lying on the ground tend to fade to a lighter color due to the bleaching effects of sunlight. Leaves scuffed up and overturned by a careless foot will show up darker than the surrounding leaves and can be seen easily, but they will eventually fade over a period of days and blend back in with the others. A man walking carefully over a carpet of leaves can do so without disturbing them, but should he be moving fast or even running, the leaves will be displaced and/or flipped over, becoming obvious to a trained eye. Tired men or men carrying heavy packs often drag their feet, churning up leaves and dead vegetation with the toe of the boot as the foot rises for the next step. Hoofed animals such as deer will rarely displace leaves except when alarmed and forced to run, but the spoor will be easily identified from the small surface area of the leaves displaced by their small, sharp feet.

As is the case with fresh grass, leaves of short, low plants or those that have been stripped off and knocked to the ground may be trodden on by the passage of human feet. If the underlying ground is stony the leaves may be crushed between the boot sole and the stones. A close examination may reveal bruising and crushing and provide conclusive evidence of human presence.

Twisted and Snagged Vegetation

When humans pass through areas of low vegetation as one would find in woods and forests, very often their passage can be determined by twisted and snagged branches and vines. In the forests of the Pacific Northwest, good indicators are ferns and brambles that become twisted and dragged out of position, thereby showing the lighter underside of their leaves. Practically all growing leaves are lighter on the underside and if exposed tend to catch the eye. Brambles with thorns catch onto clothing and are dragged along in the direction of travel. Most upturned leaves do eventually return to their normal lie within a short time, particularly if they are not damaged.

Skinned Bark

As has been pointed out, men moving through thick vegetation and brush, especially if carrying heavy packs or if tired, tend to knock into small trees and branches, causing skinning and bruising of the bark. Where bark has been skinned off, sap will accumulate and the underlying wood will show up as a lighter color and be moist to the touch. As the skinned area dries it tends to darken.

Stones and Rocks

Over time, long-standing stones and small rocks get "cemented" into the ground due to the effects of wind blowing and rain washing sand and sediment around their base. When subjected to the heat of the sun, this "bakes" the stone in place. When these stones are displaced or kicked out of position by a careless foot, an egg-shaped hole will be seen. As is the case with dead leaves, the buried part of a displaced stone appears darker, which makes it identifiable against the backdrop of lighter stones. As with leaves, this darker portion will eventually fade. The newly exposed cavity will be easy to see because of the same shadow effect we see on a fresh footprint. As time passes, leaves, dirt, and detritus get blown into the cavity, and insects take up residence. Examining the cavity will give an idea how long the stone has been displaced.

Wild animals rarely displace stones with their hooves under normal conditions, but big, clumsy human feet, particularly if the owner is tired or carrying a heavy load, do knock out stones and leave a trail. There are certain animals that search for insects and grubs under stones and rocks, however, so take care not to be fooled. Look for other indicators and confirmable evidence.

When climbing up or descending steep slopes, humans have a tendency to place their feet against embedded rocks to gain purchase. Often this causes the rocks to move and displaces the earth, which can be seen as a swelling on the lower side and freshly exposed earth on the upper. Insects dwelling under the rock will seek a new and safer abode fairly quickly, so if insects such as ants are seen in such circumstances the trail may be fresh.

Insect and Ant Nests

Ants live in every part of the United States, and their nests and hills are commonplace in the woods. Ant hills and nests that have been recently disturbed are characterized by swarms of ants scurrying around in disarray. Over time they repair any damage, but by ascertaining the rate of repair it is possible to gain a rough idea of when the damage was sustained.

Cobwebs

Spiders, as a rule, spin their webs at night

to catch unwary insects the next morning. Larger species place their webs in open places where a human, especially at night, will tend to walk. A freshly broken web, therefore, may indicate the passage of a human during the hours of daylight.

Shade Trees

Humans are creatures of comfort and will gravitate to a tree that provides good shade, especially during the heat of the day. If there is evidence that your fugitives rested there, it is possible to establish exactly when by working out the position of the shadow relative to the time of the day. The distance the shadow has moved away from the evidence can give a rough estimation of how long ago they occupied the spot. If the tree is an isolated one and the sign is all around, it will indicate that the fugitives were there for a fairly long period.

The Effects of Time and Weather on Litter

Care must be taken when assessing the age of litter found on a follow-up—unless of course it is a still-burning cigarette butt! Paper and packaging can be blown by the wind from other areas and lead you to false conclusions. Candy wrappings, cigarette butts, and paper debris quickly show the effects of the elements and start to fade, particularly on surfaces exposed to the UV rays of direct sunlight. Inks and dyes will also deteriorate quickly when exposed to sunlight, but it will take several days of constant exposure before the effects start to show, although once the exposure begins the dyes fade rapidly.

The Effects of Time and Weather on Bodily Fluids

Human excrement and urine are also types of spoor, and they are very definitely conclusive evidence of a fugitive's presence. Both can be useful in ascertaining the age of the tracks by calculating the time they were deposited.

On operational follow-ups of guerrillas during the Rhodesian bush war, we would often come upon human feces or urine stains left by a gang. In some cases the smell alone was enough to draw attention to the deposit. Insects swarming around were also a good indication.

It is possible to make a fairly good assessment of the age of feces by its condition and appearance. Soft, moist feces indicate a recent dump, but as the feces gets older a hard skin forms and it eventually dries out and shrivels up or is consumed by animals or insects.

Urine dries rapidly but will always leave a clear mark or hole on dry, sandy ground; when completely dry it will often leave a crusty deposit that can be analyzed. Insects, especially butterflies, are drawn to the ammonia content of urine—if insects are present, it will indicate that the trail is relatively fresh, certainly no more than several hours old or even less, depending on whether it is a shaded or exposed spot.

Vomit, phlegm, and sweat have all played a part in assessing the age of tracks, depending on the state of evaporation, which is influenced by ambient temperature and sunlight.

Blood Spoor

Blood drops or smears, known as blood spoor, are also conclusive evidence, especially if a fugitive is known to be wounded. As blood dries it oxidizes and changes color to a dark brown. As it crusts over, it may exhibit a shiny blackish patina. Blood can dry out in as little as 20 minutes depending on the type of blood, the amount deposited, and the weather conditions.

In almost 25 years of operational tracking I have only had to follow blood spoor twice, and on both occasions it was easier to follow the indicators than the blood itself. This is not to say that blood spoor is of no value. Indeed it can be exceedingly valuable in telling the tracker the state and condition of his quarry as well as indicating the age of the spoor. Blood spoor may be found on the ground or on vegetation, and the height of the spots or smears can give an indication as to where the fugitive is wounded and the type of wound sustained.

Arterial bleeding is usually an incapacitating and life-threatening wound and is indicated by large gobs of blood spurted out at fairly regular intervals. It is oxygenated blood, so it will be bright red in color. Expect to find the victim fairly quickly, as he will weaken and need to rest more and more frequently until he is overtaken by coma or death.

Venous bleeding is generally constant and will leave a trail or series of drops. It is a darker red than arterial blood. A venous wound generally is not life-threatening, so a victim can move long distances but will eventually slow down unless the flow of blood is stanched.

Lung-shot casualties often cough up pink frothy blood and phlegm-blood clots. Head-shot casualties are characterized by heavy, slimy, glutinous gobs of blood. Fugitive with head wounds will rarely pose a threat because most of their mental faculties will be drastically impaired by the impact trauma.

Remember that if you find red, wet blood, your quarry is close by and is a deadly threat. He is more than likely hiding and waiting for you. A cautious approach is mandatory.

LOST SPOOR PROCEDURES

As the tracker moves forward along the trail, he scans the ground ahead, moving from indicator to indicator, always recording in his mind the last spoor he can see. If, when he moves up to this last known point, he cannot see any spoor or sign ahead, he must stop. This point is known as the "last known spoor" and must be marked with a scratch on the ground, a stick, a stone, or in some other conspicuous way. When tracking as part of a team, the best way is to bring up the controller to stand just behind this point.

This brings us back to the most important rule of tracking: *never overshoot the last known spoor*. To do so will cause confusion and lead to loss of time. From a position slightly behind the last known spoor, the tracker must scan ahead to see whether he can pick up any other indicators along the most likely route. If this proves negative, he must then go into lost spoor procedures.

Every tracker, no matter how good he may be, will eventually lose the spoor at times. This should give no cause for concern because lost spoor procedures are established techniques to cope with this situation successfully. No matter how difficult or how long the lost spoor procedures take, if done correctly 90 times out of 100 you will regain the tracks. After all, a trail does not disappear into thin air . . . or does it?

Once, while on a spoor-cutting patrol, my team picked up suspicious tracks of four people moving in extended line, not a usual formation for anybody unless they are militarily trained. We quickly radioed in the coordinates and direction of travel to HQ and went off at a run on the fresh, clear spoor. Forward visibility was good, with little vegetation on the forest floor to impede vision or conceal an ambush. Soon the forest started to open up a little more, and the tracks remained visible ahead. It was obvious that the group was in a hurry, as no attempt was made to conceal or disguise the trail.

Suddenly, without warning, the tracker stopped and with a puzzled look on his face displayed the lost spoor hand signal. Closing in to discuss the situation, he informed me that the once clearly visible spoor had simply and suddenly vanished. We quickly instituted lost spoor procedures, alert to the possibility of ambush, but no matter how hard we looked there was nothing to be found. Baffled, we regrouped at the last known spoor and discussed the mystery. Another more detailed search, however, revealed several freshly shredded leaves and twigs lying on the ground.

At last, the answer dawned on us. The spoor we were following were those of a local reserve police patrol, all wearing nonregulation boots, that had been extracted from the area by a hovering helicopter. Unable to actually touch down because of the small size of the clearing, the rotor blades had severed leaves and twigs from the overhanging trees during its descent to pick up the patrol. Remember: always expect the unexpected!

Lost Spoor Procedure 1

Once the spoor is lost and cannot be seen

ahead, the first step is to mark the last known spoor (the controller can do this) and examine the most likely line of advance (see Figure 4), which should, in most cases, be a continuation of the previous direction of travel. The tracker then advances forward, keeping off the trail, looking for an indicator. He moves ahead about 30 to 40 yards examining the ground carefully. If the spoor is found, the follow-up continues as before, but should the spoor not be found on this most likely line of advance, the tracker must return to the last known spoor. He then checks out other possible lines and repeats the process. Fifty percent of the time this is all that is required to recover the spoor, but if a search of all lines proves negative, the second lost spoor procedure is implemented.

Lost Spoor Procedure 2

Using the last known spoor as the center point, the tracker moves backward along the tracks about 30 yards and commences to search in a circle (known as a 360) around the last known spoor in a attempt to "cut" the tracks (see Figure 4). The reason why the tracker moves back behind the last known spoor is because fugitive groups have been known to hook back prior to establishing a base camp or laying an ambush on the trail. If this initial search does not reveal any indicator, the tracker merely repeats the process by extending the size of the circle. Again, if the spoor is not located, the circle is extended again and again until the spoor is found and the follow-up continues.

Lost spoor procedures can be tedious and time consuming, particularly when hitting hard ground or if the fugitives are skilled in countertracking techniques (see Chapter 7, Countering the Tracker). However, even if you have to circle out to 500 yards or more, never fear: you *will* regain the spoor in 90 out of 100 instances. Lost spoor procedures can be accelerated by using the whole tracking team, which covers far more ground in a shorter time. (See Chapter 5, Team Tracking and Tactics.)

There are times, fortunately few, when even extensive lost spoor procedures do not recover the spoor. Under these conditions there are several options available to the team. The first option needs a lot of faith! Call the team together and ask each in turn what his gut feeling is about where the tracks may be heading. If the follow-up has been progressing for more than a few miles, it is possible to develop an instinctive feeling of where the fugitives may be headed. Human beings do not wander aimlessly in the woods and generally have a specific destination in mind. (Unless you are a Vietnamese immigrant searching for shiitake mushrooms! Many a search-and-rescue team has been called upon to locate recently immigrated Asians who, while searching for minor forest products, have become disoriented and fail to find the way back to their car.) By discussing this with the team and examining the map, it is likely that you will come up with the correct answer. I have done this on several occasions and it worked for me. (A word of warning—this technique does *not* work in search-and-rescue situations. Lost people often do inexplicable things and tend to wander in the most unpredictable fashion. In some cases, lost people, mainly children, have even deliberately hidden from their searchers, believing that they will be punished for getting lost.)

Another option available to the team is to assess the fugitive's general direction of travel and move forward in that direction for several miles. At a selected spot the team breaks into two sections, each one moving out 90 degrees to the original line of travel rather like the two sides of the crossbar of the letter T. Each section then cuts for spoor along their line for several miles if necessary. If the spoor is still not found, it is time for the third option: cutting for spoor.

Cutting for Spoor

Cutting for spoor is a useful technique that has been employed with great success, particularly on search-and-rescue operations. There are several instances when it can be employed in the operational role, particularly if the trail has been obliterated by rain or is several days old when the follow-up commences. In essence, it is similar to a 360,

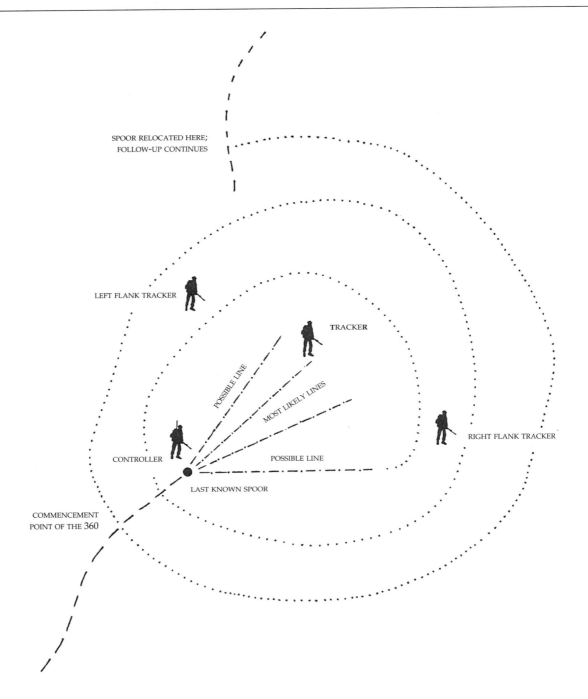

Figure 4: Lost spoor procedures 1 and 2.

1. When the tracker loses the tracks, he marks the last known spoor and moves ahead along the most likely line for about 30 yards, examining the ground. If no spoor is found, he returns to the last known spoor and repeats the process along another line. If this fails to relocate the tracks, he commences procedure 2.

2. Commencing from a point behind the last known spoor, the tracker completes a wide circle (a 360) using the last known spoor as the center point. If the spoor is not found, he widens the circle and repeats the process until the tracks are found. The flank trackers can also be used in lost spoor search procedures.

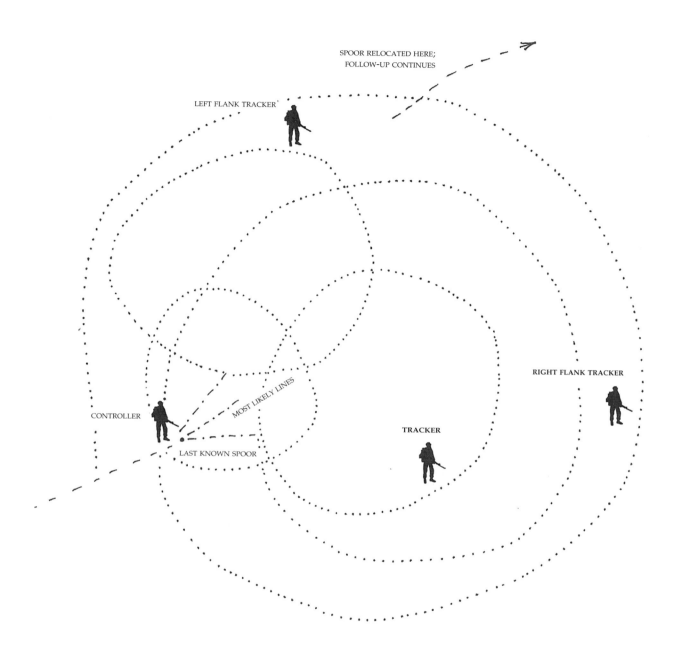

Figure 5: Extended lost spoor procedures.

If lost spoor procedure 2 fails to locate the spoor, the controller will detail the two flankers to commence further 360s out and beyond the 360 done by the tracker. The circles should intersect to ensure that the ground is covered by more than one tracker. If the spoor is still not found, more 360s are completed from the last known spoor as far as 500 yards if neccessary. While searching for lost spoor, the flank trackers are vulnerable and should scan the area carefully before moving. The controller must remain at the last known spoor while the search is going on.

but the search pattern is cast wider along predetermined boundaries.

Using a map or good local knowledge, the tracker selects several natural or man-made lines such as a road, river, forest cut, power line, or footpath that box the area where either the fugitives are believed to be or where suspect spoor has been located. These lines are searched carefully by spoor-cutting teams in an attempt to locate the spoor crossing their path. Should the spoor be found, the follow-up commences from that point onward, but if no spoor is found, it may indicate that the quarry is still inside the boundary. If this is the case, by progressively reducing the size of the box or by criss-crossing trackers through the area, the fugitives will eventually be found (see Figure 6).

A good example of the potential value of this technique occurred in early 1995 when five dangerous prisoners dug a tunnel under the fence of a correctional institution in Florida. Starting the tunnel from the basement of a prison chapel, they dug under the fence to a point several feet outside the perimeter. Upon breaking out of the ground, the first five prisoners managed to slip out unnoticed by the guards into an adjacent sugarcane field, which at that time was underwater due to flooding. The sixth unfortunate prisoner was seen by a guard while extricating himself from the hole. The guard opened fire, wounding and immobilizing the prisoner. The other five, taking advantage of the confusion, slipped deeper into the cane field and concealed themselves by immersing their bodies in the water and breathing through hollow cane stems.

Escape procedures were quickly implemented by the jail authorities, and local law enforcement was mobilized. Despite a massive manhunt, the fugitives could not be found. A week or so later, some of the escapees were rounded up and revealed that they had remained hidden in the cane field for two days waiting for the initial search activity to die down. When the net was cast wider, the fugitives had slipped out of the flooded field and were able to find places to hide and wait out the search.

The point of all this is that had trained trackers been available, they could have searched around the perimeter of the cane field and discovered that the prisoners were still inside the field. All of the search effort could then have been concentrated around the field to wait for the prisoners to give up due to cold and hunger. At the time of this book's publication, one of the jailbreakers, a violent murderer, remains at large.

BACKTRACKING

Backtracking—following the trail *back* to its source—is an important technique that can reveal much useful information. It does have several drawbacks, however. The main one is that it is expensive in manpower if another team is used to follow the forward trail. However, in law enforcement investigations such as marijuana eradication operations, it can provide a lot of clues as to where growers came from and may lead back to another grow site. In the case of a jail break, backtracking can provide critical information about the routes used by escaped prisoners, as the following story will show.

In June 1994, a prisoner escaped from a state jail farm over a long weekend. Fortunately, two members of the jail's Special Response Team had attended a tactical tracking course only a month earlier and were tasked with the follow-up. Despite being nearly four days behind the fugitive, the trackers commenced the follow-up from the point where police had located a homemade backpack. They followed the trail for several miles into the rear of a housing subdivision, where the spoor was lost due to an abundance of local movement.

Assessing that the escapee had probably vacated the area by vehicle, the trackers returned to the initial start point and began to backtrack toward the jail. Along the route they discovered where the escapee had spent the night. They recovered a powerful slingshot made in the jail and some ball bearings, which, if not deadly, could inflict a serious wound. A

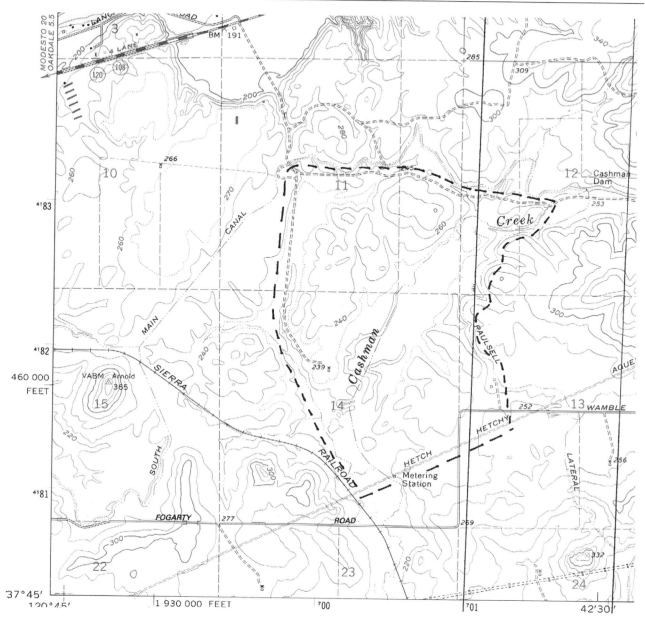

Figure 6: Cutting for spoor. Cutting for spoor is a useful technique. Choosing natural or man-made features such as roads or rivers that enclose the area, trackers check for spoor to see whether the fugitives have either exited or still remain boxed up inside these boundaries. The search lines chosen must not be too far apart but also not too close so that they don't interfere with the follow-up team on the fugitives' trail.

black plastic sheet, a pair of disposable latex gloves, and other items of jail origin were also found, positively confirming that they belonged to the fugitive.

Although the inmate was not apprehended by the trackers, mainly because of the delay in getting them on the spoor, valuable information was gained on the nature of the escape and the routes and methods used. One of the trackers, a state investigator, said to me later, "Although we were a day late and a dollar short, we were able to backtrack the route that the inmate had taken from the jail farm. We took this information back to the institution to make the best use of it we could. At a staff meeting I was able to show how a tracking team could have tracked and recovered this escapee had we been mobilized as soon as the escape had been confirmed. I am now better able to determine avenues of escape, rate, and direction of travel and approximate time of escape, all of which will benefit the department's ability to apprehend escapees and identify probable avenues of future escape attempts."

The information obtained by these trackers can now be used to construct an information database to help identify likely avenues of escape as well as provide the basis of an operational plan of action should another breakout occur. As a matter of interest, this particular institution now has an official, state-sanctioned Inmate Recovery Team made up of 18 officers who have received tracking training. Interestingly, no further escapes have occurred from this particular jail because the word has gone around among the cons that if you get out, you will be hunted down like a dog! Of course, the administration and local law-enforcement authorities are delighted, but the team is ticked off because they want to use their new tracking skills against a real fugitive! Other trained officers in the same state have tracked and recovered two escaped inmates from different jails, thus proving the value and expense of the training.

While on operations during the Rhodesian bush war, my three-man team was tasked to backtrack an interesting trail while a full team with a strong support element proceeded to track forward. The country was heavily wooded and, it being the height of the dry season, the temperature was well in excess of 100 degrees. Backtracking was easy because there were 10 people involved moving in single file along a dry sandy path. We made good time for several miles but soon ran out of light, so we moved off the trail to bed down for the night.

Early the next morning, with the sun at the optimal tracking angle, the spoor stood out like a tick on a hog's back. Being a little gung-ho we started to trot along the trail, which we judged was fairly safe to do. Our judgment (or lack of it) almost turned into a bad mistake because at around 10 o'clock, due to the rising heat of the day, we decided to stop for a short break to regain our wind. We estimated we had covered more than eight miles since departure earlier that morning.

Choosing a small rocky outcrop about 30 yards off the trail, we started to brew some tea (the tracker's favorite brew). With the water about to boil, we heard the faint sound of singing coming from the direction we were headed. Curious, we hunkered down and watched the path and eventually along came about 20 native women carrying pots on their heads, as was the local custom. We figured it was probably a feeding party returning from taking food to a group of guerrillas, because they were accompanied by a lone guerrilla carrying a Soviet RPD light machine gun.

Letting them pass, we proceeded carefully along the path, which now showed no trace of the original spoor because it had been obliterated by the women. Less than half a mile along the path we came to a large, well-camouflaged, but deserted camp with sleeping places for about 100 guerrillas. Judging from the spoor it was apparent they had only recently departed, probably several hours earlier. Trails of about 10 men each led out of the camp along different paths, so it was obvious that a large gathering had taken place the day before. Had we not stopped and brewed tea, we could have run headlong right into the well-disguised camp. Three against a hundred is certainly not good odds!

The camp was unknown to our intelligence people, so a careful watch was kept with regular high-level photo reconnaissance flights. Eventually an air strike was mounted on the base, accounting for more than 30 dead guerrillas.

As it turned out, the forward tracking team aborted its follow-up because of obviously deliberate antitracking by local natives who had overtrodden the spoor. The whole point of this story is that backtracking may lack glamor, but it certainly can pay dividends at times.

FOOTNOTE (NO PUN INTENDED!)

The Royal Canadian Mounted Police in conjunction with the University of Alberta in Calgary, Canada, recently conducted a study of criminals fleeing crime scenes and the actions taken by lost persons. Examining cases covering a period of 85 years, they have come up with some interesting conclusions that are of considerable use to trackers and indeed every police officer.

In 90 percent of the human population, the left side of the brain controls the motor functions of the right side of the body. It also controls the instinct side, which is responsible for our response to life-threatening situations (often called the "fight or flight" reflex). The right side of the brain controls the motor functions of the left side of the body and the thought process.

The study concluded that when a criminal flees a crime scene without a prepared plan of escape, the left side of the brain—the instinctive side—takes over, not only causing the suspect to run but to *make one left turn followed by two right turns*. Thereafter the pattern generally repeats itself. A fleeing suspect will go to great lengths to go to the right, including climbing an eight-foot fence on the right even if there is a three-foot fence on the left. So strong is his desire to go to the right, he will pass up possible escape routes on the left even to the point of boxing himself in.

The study also discovered that when a fleeing fugitive had a prepared plan of escape, the flight instinct still battled to overcome the prearranged plan. From a search-and-rescue perspective, this phenomena explains why lost people tend to travel in circles.

Chapter 4
CONDUCTING THE FOLLOW-UP

Boot prints are to the tracker what fingerprints are to the detective.

Charles Cates,
former Search and Rescue Coordinator,
Taos, NM

To enable a tracking team to work efficiently, certain essential elements of information (EEIs) are needed to provide the team with answers to the questions:

WHO?
WHAT?
WHERE?
WHEN?
HOW?
WHY?

Without this information trackers may waste valuable time following the wrong tracks or, even worse, blindly following the tracks of people who are desperate enough to use violence to prevent capture. As is the case in any tactical situation, to be forewarned is to be forearmed. The failure of trackers to take the time to acquaint themselves with the situation by questioning available witnesses or by examining the evidence and extracting at least a rudimentary outline of events is extreme folly and can only lead to disaster.

A tracking team does not work in isolation, and its initial assessments and conclusions obtained before commencing on the follow-up are required to brief other teams and assets involved in the operation. Good information passed back to the operations control center facilitates the correct employment of additional assets such as helicopters and fixed-wing aircraft in the most economic and effective manner.

Whether all or only some of the EEIs are necessary depends on the nature of the follow-up. The requirements of a correctional team tracking escaped convicts will differ from those of a rural sheriff's team following a band of marijuana growers as much as it will differ from a Special Force's team tracking a gang of Marxist guerrillas through the jungles of South America. Whatever the origin of the team involved, as much of this information as practically possible must be obtained before the follow-up begins.

Having said that, it must be stressed that to close the time/distance gap between trackers and fugitives, it is essential to get the follow-up underway as soon as possible, so no unnecessary time must be wasted in acquiring the EEIs. If this information is not readily available, much of it will likely become so as the follow-up progresses. The important thing is to get the follow-up going as soon as possible—if it comes to a decision to wait for the information or to get going, better to get going.

The 11 EEIs required, with the five most important in italics, are as follows:

1. *Ascertaining the correct spoor.*
2. *Assessing the number of people involved.*
3. *Recording footwear types, patterns, and approximate sizes.*
4. *Determining the initial direction of flight.*
5. *Assessing the age of the tracks (in two-hour increments).*
6. Noting any abnormalities such as wounded quarry, dogs, or bicycles, for example.
7. Getting a description of fugitives, including sex, age, clothing, packs or weapons if possible.
8. Gathering relevant information from victims or witnesses.
9. Checking history of activities and known local contacts.
10. Noting any previous records of antitracking methods used.
11. Assessing violence potential or propensity to use firearms.

ASCERTAINING THE CORRECT SPOOR

Upon arrival at the scene of a potential follow-up, trackers must correctly ascertain the spoor that they are to follow. Some situations are obvious, like the prints of escaped prisoners from a jail or from known criminals fleeing from police operations. But some scenarios can be more complicated. Take, for example, if a hiking party were attacked and robbed in a wilderness area by unknown persons and where there are a multitude of well-trodden trails used by numerous hikers. The case of the Olympic skier Kerri Swenson, who was abducted by two Montana mountain men, is an example of this. Had skilled trackers been deployed into the area to search for her trail prior to the arrival of well-meaning but essentially useless search parties, they would have found evidence of the initial encounter and been able to follow the correct spoor. It is quite likely that Miss Swenson could have been recovered before the tracks were obliterated and the trail gone cold.

Once the correct spoor has been identified, either by confirmation of witnesses, searching, or deduction, the next step can be carried out.

ASSESSING THE NUMBER OF PEOPLE INVOLVED

If witnesses are available it is no problem to find out how many people are involved, but more than likely you will have to search the area and judge the available evidence to make the assessment yourself. This is done by examining the scene, taking care not to obliterate any clues as in a normal crime scene investigation.

After all the available evidence has been gleaned, trackers should examine the most likely exit routes from the scene. In most cases this will provide you with most of the clues about numbers, direction of flight, and so on. (With a little practice it is easy to count up to five or six sets of prints if they are moving in single file.) However, if the obvious routes show no evidence of being used, the trackers then complete a 360 degree circle around the scene. By closely examining every inch of the ground, the exit route will eventually be found.

If it is determined that the group is larger than five or six, the "average pace" method is used to determine the number of tracks. This is a speedy process that should take only a minute or so. With a stick, mark off two lines across the trail 36 inches apart. Count the number of prints inside the box and divide by two. This should give you roughly the amount of people involved because each person is likely to have placed each foot within the two lines (see Figure 7). However, at this stage do not get hung up on an exact figure: experience shows that it is not that critical, and in most cases you will be wrong anyway. Far better to

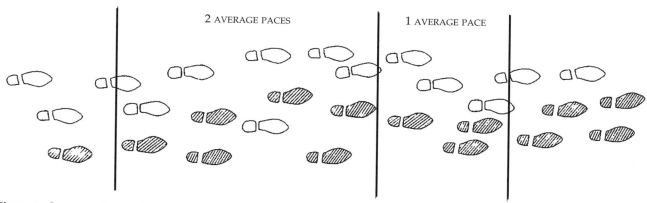

Figure 7: Counting the tracks—the average pace method.

Method 1. Calculate the average pace according to prevailing conditions. Mark the distance across the tracks with two lines, count the amount of prints within the box, and divide by two. Seven prints divided by two equals three people.

Method 2. This method is more accurate. Repeat as above but measure off a two-pace distance instead of one and mark. Count prints in the box and divide by four. Twelve prints divided by four equals three people.

say something like "between 10 and 15" and get on with the follow-up. A correction can be made later when more information is obtained.

RECORDING SPOOR PATTERNS

Because it is essential to pass useful information on to the operations control center and other tracking teams, it is important to note and record footwear patterns worn by the quarry. This is not an easy matter because there are thousands of different patterns available in the United States, and most of them, especially athletic shoes, have complex designs. A Polaroid camera is useful for this if prints are clear and photographable.

Search and rescue teams throughout the United States have adopted a system that classifies footwear into two basic groups: heels and flats. Heels refers to footwear with a separate and distinct heel, such as formal shoes or boots; flats refers to a heel-less shoe, such as basketball or athletic shoes used by runners.

Some spoor patterns such as Vibrams, Panama soles, and waffles are easy to describe and are known to most people, but others are extremely complicated, particularly athletic shoes. Examine a pair and you will see what I mean. Trackers should practice describing footwear patterns over the air to other trackers until a good level of competence is attained.

An excellent technique is to use a spoor card. This is a card, or series of cards, approximately six by four inches, divided into 12 squares. Each square contains a diagram of a footwear pattern in common use, and each square is coded. If information on spoor patterns has to be sent over a radio, the operator merely refers to the code that best describes the relevant spoor pattern. For example "I have two type X and three type Y." This system simplifies communications considerably and is quick and easy to understand.

DETERMINING THE INITIAL DIRECTION OF FLIGHT

When commencing a follow-up it is important to ascertain the initial direction of flight so that your HQ can plot the information on a "going map." Although the direction taken by a fugitive group can and will change from time to time, this information assists the command-and-control element in formulating countermeasures such as siting observation posts, laying ambushes, and directing spoor-cutting patrols. It is not necessary to send an azimuth when reporting direction of flight—the cardinal points of the compass are used.

While on the follow-up, any major change in direction must be relayed back to your HQ so that the trail can be plotted on the going map and an accurate record of the follow-up can be maintained. A global positioning system (GPS) is invaluable, as it will record exactly where the follow-up commenced and each and every change of direction as it progresses.

ASSESSING THE AGE OF SPOOR

When commencing a follow-up, knowing the time the spoor was laid is invaluable for two reasons: (1) it will indicate how far the group is likely to have traveled so an assessment of their present whereabouts can be made; and (2) it indicates how much time and distance the tracking team has to make up to make contact with the fugitives.

There are a few ways to assess the age of spoor:

- By knowing the exact time of the incident (for example, the escape of a group of prisoners from a jail work party)
- By asking witnesses who can pinpoint the exact time
- By making an assessment based on skill, experience, affects of the weather, and other factors

Irrespective of the method used, do not waste precious time looking for witnesses. The important thing is to get the follow-up underway as soon as possible. Once the follow-up is in motion, missing pieces of the puzzle often become apparent and soon you will have a clear picture of exactly who, what, where, when, and how in respect to the situation.

No matter how many follow-ups you may engage in, you will always be in for a surprise, as the following story illustrates.

During the Rhodesian bush war, my HQ received a radio message from an observation post (OP) sited on a hilltop that an explosion had occurred in a densely wooded area close by. Having observed the smoke from the blast, the OP calculated the grid reference of the position and a tracking team was deployed by helicopter to investigate. After finding the site of the explosion, which was obvious from the shredded clothing and pieces of flesh and bone, the team leader reported blood spoor and tracks of one person leading away from what appeared to be the remnants of a booby trapped radio set of common Japanese manufacture. A pack and a blood-smeared AK-47 rifle were found, which confirmed the tracks to be of one of the bad guys.

A short follow-up ensued and within two hours the team, quickly closing the time/distance gap, was amazed to observe not one but two terrorists, one being carried on the back of the other. Realizing that the terrorists were unarmed and incapable of resistance, the trackers surrounded and captured the twosome. One, blinded by the explosion, was carrying the other, who had both legs shattered by the same blast. The blind one provided the leg power and the legless individual provided the eyesight. So impressed were the tracking team by the determination and courage displayed that they provided the pair with rations and water and organized medical treatment and evacuation. In the subsequent debrief it was discovered that the pair had moved three and a half miles from the site of the explosion in less than two hours! If you still haven't realized what an amazing feat of endurance this is, try carrying one of your buddies of similar size and weight with your eyes closed for just 200 yards and see for yourself how exhausting it can be.

The point of this story is: never assume things are what they appear to be. *Always expect the unexpected.*

ANY ABNORMALITIES

Information on any abnormalities is important to trackers when engaged on a follow-up. The presence of wounded fugitives, for example, will suggest that blood spoor is a possibility or that a severely wounded fugitive may have to be left behind rather than slow down the escaping group. Bicycles and dogs leave distinctive spoor patterns and may be crucial in identifying the correct spoor later on, especially if other teams are utilized.

DESCRIPTION OF FUGITIVES

Although it may not be possible to obtain this information unless witnesses are available, anything that may physically identify the fugitives is of great value to the tracker. Information on sex and age may indicate an ability (or inability) to cover distance, especially over difficult terrain. Descriptions of clothing can assist in identification from observation posts, chance observers, or airborne searchers. (During the Rhodesian bush war, terrorists habitually wore several layers of clothing so that if seen and pursued, they could change their appearance by shedding the outer layer of clothes in an attempt to confuse their pursuers.) Whether fugitives are carrying packs is important in assessing their ability to move fast or their susceptibility to tiring quickly. Any information on whether the fugitives are armed or not is of course crucial to the safety of the follow-up team and in the formulation of the correct tactical plan.

INFORMATION FROM VICTIMS OR WITNESSES

If they are available, victims or witnesses can provide crucial information to the tracker. Apart from physical descriptions as outlined above, intentions or future destinations may be known or have been overheard by witnesses or victims and may be invaluable in planning exactly where to use scarce law enforcement resources.

HISTORY OF ACTIVITIES AND KNOWN LOCAL CONTACTS

Any knowledge concerning the past activities of the fugitives will be of value in assessing their possible destination. Trails and areas frequented by the fugitives may provide clues as to where to employ spoor-cutting teams. Any local contacts, if known, can be kept under surveillance. This applies particularly to teams involved in marijuana eradication programs in remote, heavily wooded rural or mountainous districts.

RECORDS OF ANTITRACKING METHODS USED

Any tracker wants to know exactly what he is up against and the capabilities of his quarry. If trackers have been used in the area before, it is likely that antitracking methods may have been devised, especially in marijuana growing areas where cultivators are sensitive to law enforcement attention. Knowledge of these measures may result in the prevention of casualties among the tracking team, particularly if plants, fields, or paths are known to have been booby-trapped in the past. (See Chapter 7, Countering the Tracker.)

ASSESSMENT OF VIOLENCE AND USE OF FIREARMS

Obviously, any follow-up team will want to know the propensity for or the likelihood of violence on the part of the quarry. This knowledge will dictate what weapons and equipment the team will have to carry and the type of tactics to employ to counter the threat posed. As with other law enforcement personnel, trackers must always consider their personal safety while engaged on a follow-up.

SENDING A SITREP

You have arrived on the scene, examined it carefully but quickly, and established the EEIs. You now know how many people are involved, you have recorded the spoor patterns, you know the direction of flight, and you have established the age of the spoor within a reasonable margin of error. The following steps must now be carried out prior to the follow-up getting under way:

1. Hold a quick team conference to ascertain that each member is aware of all the information relevant to the follow-up and the tactical plan advocated by the controller.
2. Outline the tactical situation, allocate team positions, and any brief support elements.
3. Send a situation report (SITREP) by radio back to his HQ containing information on the location, number of tracks, direction of movement, age of the tracks, and type of footwear with a description of the sole patterns. This is known as a LNDAT which is an acronym for the following:

- *Location*: your current position
- *Number*: number of individual sets of tracks
- *Direction*: the initial direction of flight by the fugitives, sent in compass direction
- *Age*: the approximate age of the tracks in two-hour increments
- *Type*: type of footwear patterns

It is not necessary to give the main headings each time a SITREP is sent. To speed the radio transmission, phonetic alphabet headings are used. Mandatory headings are as follows:

- ALPHA (A) means location.
- BRAVO (B) means number of tracks.
- CHARLIE (C) means direction of flight.
- DELTA (D) means age of the tracks.
- ECHO (E) description of tracks.

Operational headings are covered by:

- FOXTROT (F), which is anything else relevant to the follow-up

Therefore, a typical opening LNDAT SITREP could sound like this:

- ALPHA Grid 241680
- BRAVO 4
- CHARLIE Southwest
- DELTA 12 hours
- ECHO 2 Redwing boot, 1 cowboy plain sole, 1 Vibram sole
- FOXTROT Blood spoor seen

Nothing more is necessary. Get on the radio, pass the message, and get on the trail without wasting any more time.

SITREPs using the above format are periodically sent back to HQ to enable the command staff to plot progress of the follow-up on the map board and plan appropriate countermeasures. (See Chapter 8, Command and Control of a Tracking Team.)

The foregoing activities seem to be a lot but in actuality take only a few minutes to complete. The more follow-ups you do, the better you will become at them.

Chapter 5
TEAM TRACKING AND TACTICS

While one man alone may observe certain tactical principles, any study of tactics must deal with the cooperative efforts of a team.

Lt. Col. Jeff Cooper

Unlike search-and-rescue tracking, where it is normal to use only one tracker, the hunt for dangerous fugitives must be conducted in an entirely different way. When faced with the possibility of firearms being used against you, you have to think in terms of tactics and protection. The only way to do that successfully is with a tactical tracking team.

Although I have conducted several follow-ups against armed insurgents as a solo tracker due to operational requirements, it was never a pleasant experience, and it was undertaken only as a *last* resort. The mental strain under such circumstances eventually creates a counterproductive situation in which the tracker does not perform to his optimal capacity.

ADVANTAGES OF A TRACKING TEAM

A well-trained tracking team that has worked together for a period of time can reach a level of proficiency that far outdistances even the most competent tracker working alone. A team effort has a number of advantages:

1. A team is self-contained and provides its own protection on a follow-up.
2. All members of a team are cross-trained in all team roles and positions so they can be rotated while on a follow-up to prevent fatigue.
3. A team can track faster and relocate lost spoor more efficiently.
4. A team has the ability to gather information faster and utilize it effectively.
5. A team is versatile and can be used in a number of other roles besides tracking.

Just about all SRTs, be they police, corrections, or military, consist of between 8 and 24 members, so establishing a tracking team should present no problems. The most effective tracking team, based on considerable experience, consists of four trained individuals, although three may be sufficient if there is no armed threat to the team. The team of four has proved to be the most suitable for long-term follow-ups due to its flexibility, mobility, and firepower.

A team consists of four distinct roles, but all

Lightly equipped, fast-moving Rhodesian Army trackers prepare to be uplifted by helicopter to investigate guerrilla tracks found by a police patrol during the bush war. It was army policy to send trackers out to investigate every guerrilla action, sighting, or even suspicion of terrorist presence—a policy that paid handsome dividends.

members should be cross-trained and capable of assuming each role when required. These roles are described here.

The Controller

The controller is responsible for the tactical movement of the team. He carries the radio and is responsible for communications both with HQ and any support elements involved, including direct communications with air assets. He need not be the most senior or experienced member, but he has to have an excellent grasp of follow-up dynamics and tactical movement. He is responsible for maintaining visual contact with the tracker and the flank trackers and for all tactical formations and operational decisions. He is responsible for marking the last known spoor when requested to do so by the tracker.

The Tracker

The tracker's only task is to follow the spoor and, based on his interpretation of the indicators, evaluate the likely actions of the quarry on an ongoing basis. He informs the controller of his evaluations and assumptions. He must have excellent eyesight and a high level of patience.

Left Flank Tracker

The left flank tracker is responsible for the security of the tracker by protecting the left flank of the formation. He is responsible for scanning his area of primary responsibility for ambushes, obstacles, booby traps, changes in terrain, and anything else that may affect the tracker or distract him from his task. He also searches for lost spoor under the direction of the controller. He must be prepared to take over as tracker any time the tactical situation demands.

Right Flank Tracker

The right flank tracker is responsible for the right flank in the same way the left flank tracker is for the left flank.

The guerrilla tracks referred to in the previous photo led to a buried arms cache containing 3,000 rounds of ammunition for the AK-47, eight B40 antitank rockets, and three TM47 antitank mines, one of which could immobilize a five-ton truck, armored car, or tank.

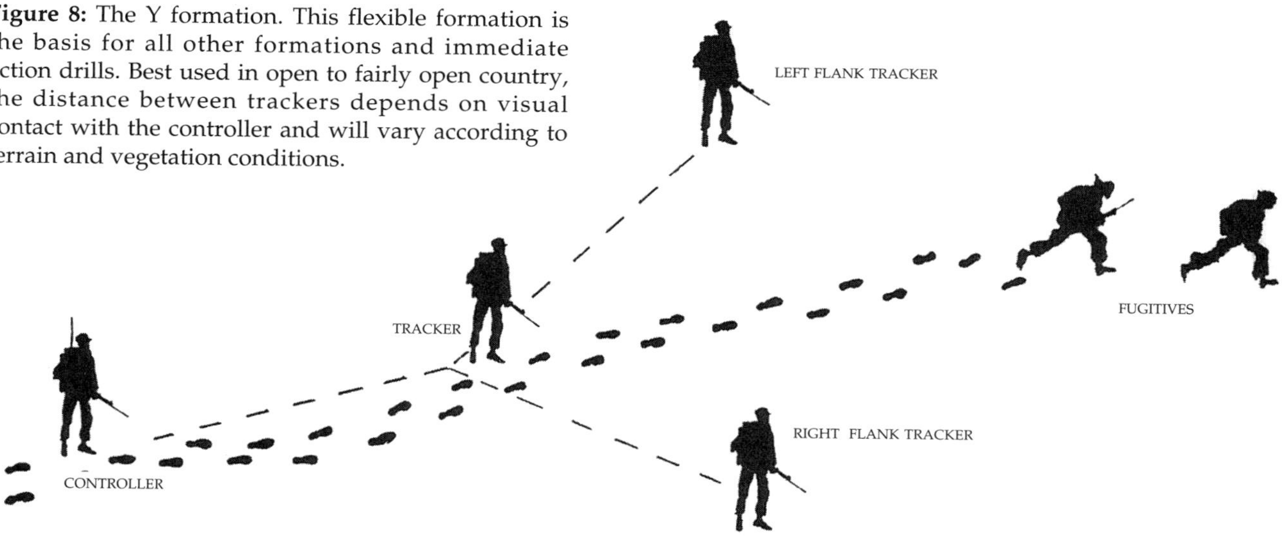

Figure 8: The Y formation. This flexible formation is the basis for all other formations and immediate action drills. Best used in open to fairly open country, the distance between trackers depends on visual contact with the controller and will vary according to terrain and vegetation conditions.

TEAM FORMATIONS

The basic tactical tracking formation is the Y formation, which represents the position of the team on the ground. The controller is at the base of the Y, the tracker at the fork and the two flankers at the ends of the arms (see Figures 8 and 9).

From this basic formation, the team can quickly reconfigure to several other patterns according to ground and vegetation conditions.

These other patterns are the half-Y (left or right), single file, and extended line (see Figures 10 and 11). These formations have been found to be all that are required for practically all tactical tracking operations

The Y Formation

The Y formation is generally used in open or lightly wooded areas. The size of the formation is dictated by the visibility between the controller and the rest of the team. Should the

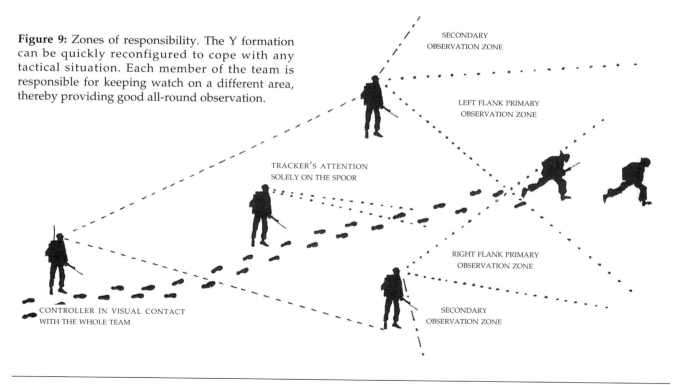

Figure 9: Zones of responsibility. The Y formation can be quickly reconfigured to cope with any tactical situation. Each member of the team is responsible for keeping watch on a different area, thereby providing good all-round observation.

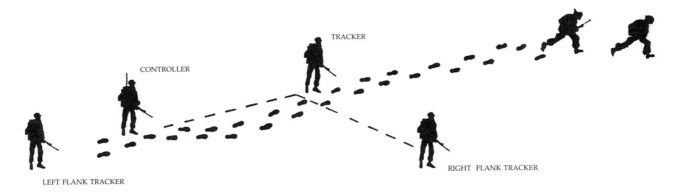

Figure 10: The half-Y formation. The half-Y is used when ground or vegetation conditions are such that the flank tracker cannot keep up with the speed of the follow-up. Although he is now positioned to the rear of the controller, he is still responsible for observing his zone of responsibility.

country be open, the team can spread out to about a 30 yard front, but if the country becomes thickly vegetated, it must shrink in size based on the team's ability to retain visual contact with each other. An additional advantage of this formation is that if the trail should swing away from the direction of travel, the flank tracker will pick it up before the tracker reaches that spot. Similarly, if the fugitive's tracks suddenly turn or break away, the flank tracker should pick them up first. The flanker can either call the tracker forward or take the spoor himself, while the tracker moves out to the vacated flank position. In this way, no time is lost and the rotation of trackers prevents fatigue on a long follow-up.

When tracking conditions are difficult due to hard ground, antitracking measures, or bad light and the spoor is temporarily lost, all three trackers can search for the lost spoor, which saves times and results in wider search patterns. On several follow-ups in Rhodesia's wildlife areas, spoor was lost among numerous animal tracks; the 360s often extended out for up to a half-mile before the spoor was relocated.

The Half-Y Formation

Should the fugitive's trail run, for example, along a riverbank, it is impractical to continue to use the full-Y formation because it would mean that one of the flank trackers would have to cross the river to maintain the formation. Far better to bring him in to provide close-in protection for the tracker while maintaining observation of the opposite riverbank (which is a good place to site an ambush). Similarly, if the trail should follow along the base of a steep hillside, to place the flank tracker on the slope would reduce the speed of the team. It is better to bring him in with the tracker to maintain the momentum of the follow-up. In any situation where one of the flank trackers has to traverse thick vegetation or broken ground, it is advisable to bring him in except when the tracker assesses the spoor as less than 30

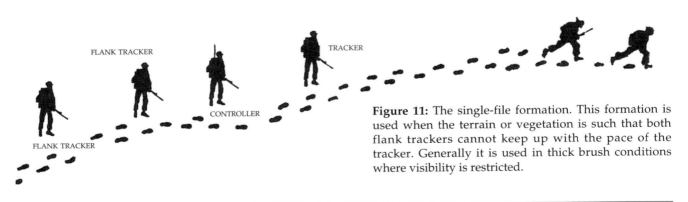

Figure 11: The single-file formation. This formation is used when the terrain or vegetation is such that both flank trackers cannot keep up with the pace of the tracker. Generally it is used in thick brush conditions where visibility is restricted.

minutes old, at which time normal tactical awareness must be implemented.

It must be remembered that the aim of a tracking team is to cut down the time/distance gap so the fugitives can be apprehended. The speed of the follow-up must be dictated by the tracker and not by the flanks. The controller must be constantly alert to ground and vegetation changes and alter the configuration of the team to maintain the best possible speed.

The Single-File Formation

When the team encounters thick brush and visibility is limited, the single-file formation is used. The tracker remains on the spoor protected by one of the flank trackers. The controller places himself in a position where he can maintain visual contact with the tracker. The other flank tracker stays to the rear of the controller, ready to move forward to his normal position should the vegetation thin out. Again, trackers must remain in visual contact with other team members at all times.

Extended Line Formation

This formation is adopted when the country is completely open and there is obviously no threat from ambush by the fugitives. It is also useful when the spoor is hard to see and all team members can track simultaneously. The best distance is 10 to 30 feet between trackers so that a front of between 40 to 100 feet is covered.

TEAM COMMUNICATIONS

On a follow-up where there is the possibility of physical danger to the team, it is essential that communication between the members be kept at 100 percent at all times. As stated earlier, the team basically relies on visual contact, which is silent and requires no unnecessary noise that could warn the fugitives of the team's approach. It is the flank trackers' responsibility to look to the controller regularly for instructions, especially when conditions start to change. In this way, the

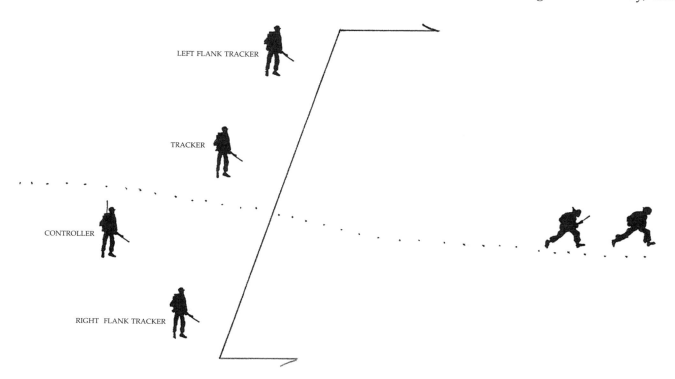

Figure 12: Extended line formation. The extended line is used when there is little vegetation to restrict visibility and the spoor is difficult to see. Using three trackers simultaneously speeds up the follow-up and keeps momentum going.

controller can pass on any instructions by use of silent signals. It does not take a team long to "click" together and develop what appears to be a telepathic communication system.

I once had a tracker who was one of the most intuitive people I have ever met. He just seemed to be exactly where I needed him. All I would have to do was look and he would be right there. This is a good argument for keeping the same team together—it has advantages that are lacking in new or ad hoc teams. Only by constantly working together do these special skills evolve.

In the early days of tracking operations in Rhodesia, our teams used a silent dog whistle employed in K9 training. When blown into in a spitting way, it produced a sound similar to a local common beetle. To the untrained ear the noise had no human connection and was thus meaningless, but to the trained ear it was as noticeable as a referee's whistle in a football game. When he wanted the attention of his team, the controller would sound off. Upon hearing the signal the team would look in for directions.

Voice-activated throat mikes and headsets configured like hearing aids, which are in common use with SRTs nationwide, are useful additions to the tracker's inventory. However, note that care must be taken not to interfere with the tracker's natural hearing ability. Radio sets should be set on squelch; it's tougher on the batteries but better for the trackers. A simple click code can be devised to communicate instructions and prevent unnecessary talking over the air, which could give away the presence of the team.

Silent Signals

Silent signals impart information and orders without using the voice or radio. There are many instances when the controller may wish to pass instructions of a routine nature to his team members without using his voice or the radio. Indeed, it is possible to conduct an entire follow-up without saying a word. Most police officers already have a rudimentary knowledge of military-type silent signals, but trackers have to know certain signals peculiar to their occupation. The main ones are described here.

Spoor Lost

This signal is used by the tracker to indicate to the controller that he has temporarily lost the spoor. It is shown by holding out the hand with the palm facing forward and the fingers extended and pointing directly upward. The controller must then stop the follow-up and stand next to the last known spoor while the tracker attempts to relocate the lost tracks.

Spoor Found

This signal indicates that the tracker has relocated the spoor. The signal is the reverse of spoor lost—the hand, with the fingers extended, is pointed down toward the relocated spoor. The controller then informs the flank trackers and the follow-up continues. This signal is also used by the flank trackers to indicate to the controller that they have found spoor.

Lost Spoor Search

When the tracker indicates to the controller that he cannot find the spoor after completing his lost spoor procedures, the controller must inform the flank trackers. If he wishes to use them to assist in relocating the tracks, he does so by pointing his hand index finger upward, revolving the hand, and pointing in the direction he wants searched.

Fugitives Seen

A team member who spots the fugitives immediately freezes, then slowly points his weapon in their direction while maintaining an aiming posture. The rest of the team then looks to where the fugitives are. When they identify the fugitives, they in turn signal this by raising their weapons to the shoulder. The controller then issues instructions or the team goes into an immediate action drill as dictated by the situation.

Other silent signals can be devised by teams

to suit their particular requirements, but the four outlined above are essential to tracking teams and have become standard worldwide.

WORKING WITH SUPPORT TEAMS

Tracking teams generally work alone, but if the fugitives are well armed the assistance of support teams will be necessary. Support teams follow the tracking team at a tactical distance and supply the muscle if contact is made with the fugitives and a major firefight ensues. Depending on the size and armaments of the fugitive group, the support troops could be squad, platoon, or company size. In a military context, these troops are generally infantry, but in the case of police or corrections, they can be other members of the SRT or normal duty officers.

During the early days in Rhodesia, we were escorted by several soldiers who accompanied the flank trackers to provide additional firepower if required. Rather than help, however, they were a nuisance—they lacked spoor awareness and noise discipline—so they were quietly discarded. Trackers prefer to rely on their own skill and stealth to detect fugitives and not to draw attention to themselves by advertising their presence with noisy escorts.

If support teams are used, they are responsible for their own tactics and protection. They follow the trackers at a distance determined by visibility. As the country opens up, the distance can increase, but as visibility is reduced they must close in. Support forces must always

- be in radio contact with the controller,
- be in a position to support the tracking team if a surprise contact develops,
- remain silent, and
- never slow down the tracking team.

In practice, most support groups are not as physically fit as tracking teams and tend to hold up progress, particularly in the afternoon and early evenings when tracking conditions are optimal. This is not a criticism, but they always seem to carry too much equipment and tire faster. Our teams would prefer to track alone and rely on the mobile helicopter-borne fireforces if they got into any trouble.

My three-man team was called in to assist a police antiterrorist unit after a farm was attacked by terrorists. Upon arrival at the scene we quickly went through the essential elements of information, briefed the police commander, and set off on the tracks of five terrorists. The tracking was fairly easy and we moved at a fast pace. About an hour later we came to a suspicious area with a large amount of spoor going in and out of a rocky riverbed. Wanting to brief the police officers on the support team, we waited and waited for them to catch up. After repeatedly trying to raise them on the radio, we gave up and proceeded to search the riverbed ourselves. Following the spoor on the soft sand, we quickly discovered several hastily covered holes and unearthed a large number of packs, weapons, mines, and antipersonnel grenades.

As it was getting dark, I again attempted to reach the police HQ on the radio and eventually managed to do so. It turned out that the support group had returned to base because they said we were going too fast for them to keep up. To add insult to injury, we

Rhodesian Army and police tracking teams were successful in tracking down and eliminating armed insurgents. These weapons—three AK-47 rifles and four SKS carbines—were recovered after a contact with a group of eight guerrillas. Only one guerrilla escaped.

were ordered to remain and ambush the area overnight. Without sleeping gear, the three of us sat huddled together for what turned out to be the coldest night in more than 50 years! Needless to say, our relations with the police in that district were strained, but they did try to make it up to us with a plentiful supply of food and beer the next night.

The point is, never let your support group slow you down. Had we slowed up in the above case, we might not have found the cache and denied the terrorists their weapons and equipment. It is the responsibility of the support troops to keep up. They are there to support, not hinder, the follow-up.

IMMEDIATE-ACTION DRILLS

Immediate-action drills (IADs) are a set of rehearsed actions taken by a tracking team to counter most tactical problems they may encounter while on a follow-up. These situations include

- when the team sees the fugitives first,
- when both the team and the fugitives see each other simultaneously, and
- when the team is ambushed.

There are many possible situations, especially during a follow-up, where the need for IADs are necessary. Teams need to practice these drills until they become second nature—they are not called immediate action drills for nothing. In the case of armed and dangerous fugitives, well-rehearsed reactive drills may be critical in ensuring officer safety and survival.

If the follow-up is supported by troops or other officers, the controller has the responsibility of briefing them prior to commencing the follow-up. Subjects covered include the actions to take in the event of the tracking team's encountering the fugitives unexpectedly as well as a description of what IADs the team will take under the various circumstances described. The controller must ensure that all radio sets are on a common frequency and communications are tested before moving out on the follow-up.

Should the controller have any doubts about the ability of the support element to keep up with and effectively support the tracking team in a firefight, he has no other option but to tell them he has no use for their services and politely advise them to return whence they came.

When the Team Sees the Fugitives First

When a member of the team sees the fugitives without having been seen himself, the IAD is to freeze and slowly bring his weapon up into the firing position pointed in the direction of the fugitives. This has two advantages. First, it indicates to the rest of the team the position of the fugitives, and second, should the fugitives become aware of the team, at least one member can open fire if necessary. In the frozen position, the team member in visual contact with the fugitives watches to see exactly what they are doing. The fugitives may be either stationary, or they may be moving toward, away from, or obliquely across the front of the team.

Depending on the distance between the trackers and the fugitives, the controller then decides what the next course of action will be.

If Fugitives are Close (Within 50 Feet)

Depending on his agency's rules of engagement, the team member closest to the fugitives will announce himself in a loud and commanding voice and instruct the fugitives to drop their weapons and raise their hands. If the fugitives comply with the order, the normal disarming and arrest procedures take place. Should the fugitives raise their weapons or attempt to fire, the team members, who already have their weapons trained in the ready position, will open fire. In the case of a military team, this drill becomes an immediate ambush and they will open fire at the most opportune time to inflict the most casualties upon the enemy.

*If Fugitives are Stationary,
but Some Distance Away*

Depending on the distance and taking into consideration the vegetation and the nature of the ground to be covered, the controller will carefully move his team forward. While on the move, the team members will keep their weapons pointing toward the fugitives so they control the initiative at all times. When close enough, the controller will announce himself and challenge the fugitives to drop their weapons and raise their hands in the same way as described above.

If Fugitives are Moving Away from Team

Should the fugitives be unaware of the presence of the team and be on the move, the controller must order his team to move up rapidly but silently and maneuver them into a position where the fugitives can be challenged in the normal way. Care must be taken so that if the fugitives become aware of what is happening and attempt to flee, they can be apprehended immediately. Should they manage to escape, the follow-up will have to be restarted.

When Team and Fugitives Sight Each Other Simultaneously

This situation occurs frequently on follow-ups, and the tracking team must take immediate control of the situation by advancing aggressively toward the fugitives with their weapons raised in the firing position. The controller must challenge the fugitives to drop their weapons and raise their hands. It is important for the team to position itself so that none of the fugitives are tempted to flee. If the fugitives raise their weapons, the team will already be in a position to initiate fire immediately.

In a military situation, the team must dominate the situation and place accurate fire upon the fugitives to inflict maximum casualties in the shortest time. If support troops are available, the tracking team should try to pin down the fugitives until the support element can maneuver to assault the enemy position.

When the Team is Ambushed

Should the tracking team ever be ambushed by the fugitives it is tracking, then somewhere, somehow, the team was not doing its job properly. However, this has happened before and very likely will happen again.

Experience has proven that the most successful method of countering an ambush is to charge right into the ambush position itself. This is the least likely course of action expected by the ambushers and has the double benefit of closing with the enemy and breaking out of the trap. To remain in a well-chosen killing ground is suicidal, especially if the ambush site has been chosen skillfully. To try to exit the killing ground after the ambush has been sprung in any way other than *through* the ambush itself runs the risk of encountering booby traps and other exotic devices deliberately designed to spoil your day.

RUNNING ON THE SPOOR

There are times when running on the spoor is permissible. If the conditions are right and it is safe to do, it can be an excellent way to cut down the time/distance gap. If the spoor is clear, visibility ahead is good, and the spoor's age is assessed as more than 12 hours, running on the spoor may be considered safe to do, but with the proviso that as soon as conditions change or the sixth sense starts nagging, you must revert to a normal, safe, tactical pace.

Running on the spoor does not mean a blind, headlong rush into the unknown. Indeed, all factors must be considered before embarking on this type of follow-up technique. There are several negatives that must be considered:

- Noise discipline is compromised. No matter how well your equipment is tied down, it is inevitable that the noise level will rise with sloshing water bottles, creaking equipment, and especially the pounding of running feet.
- If the trackers are not supremely fit they will tire, which could lead to other problems later on.
- Some vital indicators may be missed.

- Water consumption increases dramatically.
- Rest and recovery times are longer.
- The possibility of an ambush is increased.
- There is a distinct possibility that the tracker may overshoot the last known spoor.

If you do decide to run on the spoor, you must weigh up all the factors both good and bad before you come to a decision. Trackers involved in search-and-rescue operations should also consider the above nontactical factors before they decide to run.

One follow-up in which I was intimately involved turned into a disaster because all the above factors were not considered. It was a case of natural, youthful aggression overcoming caution. Tracks of about 30 guerrillas were found by a dawn police patrol, and as luck would have it, I happened to be the commander of the fireforce that day. We were called out and a follow-up commenced.

I had three sticks of four men each, fortunately most of them trained trackers. (In Rhodesia, the smallest army subunit was a "stick" which consisted of four men armed with one light machine gun and three FN FAL rifles.) I put one stick on the spoor and, not wanting to distract them with the beat of turbine blades, dropped the other sticks down on a nearby hilltop while the choppers went to refuel and bring in more troops. With the spoor barely two hours old and closing, the tracker on the spoor, Corp. Doug Cookson, informed me that the gang members knew they were being followed and were herding cattle behind them to conceal their tracks, and this was slowing down the follow-up. I uplifted another team to assist

Trackers must make every effort to cut down the time/distance gap between themselves and their quarry. Using helicopters to drop trackers on the spoor saves time and can be used to leapfrog teams forward to investigate likely escape routes or hiding places.

Cookson and, with eight trackers working on a wide front, they began to make better progress.

A few minutes later I was informed by my chopper pilot that the red warning light had come on and it was time to refuel. Handing over control to my commanding officer who had just arrived in another gunship, we returned to base about 25 minutes flying time away. Grabbing a quick sandwich as the chopper was refueled, I tuned into the fireforce frequency on the radio. Within minutes the electrifying words "Contact, Contact, Contact" came in loud and clear over the air, and I could clearly hear a tremendous amount of firing and explosions in the background of the radio transmissions. Refueling completed, we took off in record time and, skimming the treetops at 150 miles an hour, returned to the contact area.

On the way back my C.O. briefed me by radio on the situation on the ground. The tracking team so ably led by Doug Cookson, a highly motivated and aggressive soldier, had run right into a hastily prepared but effective ambush. Cookson was dead, an officer and corporal were wounded and still lying in the killing ground, and a fierce battle raged around them. Of the original eight men on the spoor, one was dead and three wounded, two seriously.

With other support troops still about 20 minutes out, I uplifted the sticks that I had dropped earlier and redeployed them into a small clearing slightly to the rear of the four trackers who were still in close combat. With my C.O. acting as an airborne commander in the other gunship, I landed, organized the troops on the ground into a ragged sweep line, and advanced toward the firefight. The ground was exceedingly rocky, with thick, lush vegetation reducing visibility to only a few feet. As we joined the embattled trackers the guerrillas broke and ran in all directions. Scattered, almost hand-to-hand battles continued throughout the day in the junglelike conditions. The overhead canopy was so thick that the circling gunships and rocket-armed aircraft were unable to acquire targets and assist us with supporting fire.

Shortly before last light, dirty and tired, we were able to regroup and count the cost: 29 guerrillas dead and two captured. As well as Doug Cookson, Sgt. Peter White, a decorated hero and one of the best soldiers I ever had the privilege of working with, Sgt. Richie Smith on attachment from the Tracker Combat Unit (TCU), and another trooper had been killed. The wounded officer and corporal both recovered from their wounds to soldier on. We lost three first-rate trackers that day, killed by an enemy that had fought with uncharacteristic courage and tenacity. It is to those three trackers, and several others, that this book is dedicated.

The lesson: running on the spoor can and does cut down the time/distance gap, but if used against armed and dangerous fugitives, dash must not supersede discretion.

OTHER TEAM TASKS

The size and flexibility of the four-man tracking team combined with its specialized training make it eminently suitable for a variety of other operational tasks commensurate with law enforcement goals. Some of these tasks are the following.

Observation and Surveillance

With their understanding of antitracking and silent movement, a team can be inserted clandestinely in rural and remote wilderness areas to observe criminal activities. I am not talking about the normal surveillance undertaken by regular officers but times where stealth and skill are required to penetrate and locate criminal enterprises such as marijuana growing operations or clandestine meth labs. A tracking team working on marijuana eradication tasks, for example, has the ability to criss-cross an area where growing activities are thought to be taking place, locate spoor, follow them to the source, and mount a 24-hour watch on the area. Much information regarding the growers and grow sites can be obtained that may prove invaluable in obtaining arrests and convictions.

As is well known, marijuana cultivators are

becoming increasingly sophisticated in concealing their irrigation systems, and the booby trapping of fields is an ever-present hazard for officers. It is only a matter of time before an officer falls prey to an explosive device set up to protect growing plants.

Other surveillance tasks include investigation of timber thefts, cattle rustling, and illegal dumping of toxic waste, as well as the observation of right wing hate organizations, survivalists, or other groups conducting tactical or firearms training. Another is checking on border crossing places used by illegal immigrants, dope dealers, and gun runners.

During a training program that I conducted in Rhodesia, I tasked several teams with observing and reporting activities at various unsuspecting sites. One team had to infiltrate and observe a tsetse fly control point where motor vehicles were examined and sprayed to prevent the deadly fly from spreading into cattle-producing areas. After a successful night infiltration, the team concealed itself in a large, thick, green bush about 30 feet back from the road, which gave them clear visibility of the control point about 400 yards away.

After several hours of normal routine activity, they observed a pink Volkswagen Beetle exit the checkpoint and drive toward them. Instead of continuing down the road, the car stopped directly in front of the bush concealing the team. An attractive young lady exited from the vehicle and walked down to the bush, whereupon she slipped down her shorts and proceeded to relieve herself about six feet from the team. Unable to restrain himself, one of the trackers whispered in a voice loud enough for her to hear, "I can see you." She leapt about four feet into the air, spun round, and in one bound, with her shorts still around her ankles, dived into the Volkswagen, which sped off in a cloud of dust!

Listening Posts

Many Rhodesian trackers, coming from rural and farming backgrounds, could speak the local native languages fluently, so we were often tasked with infiltrating villages at night and listening to the locals gossiping about recent events in the district. Very often these listening groups discovered the presence of terrorist groups in the area, which led to later operational successes.

Sgt. Russell Williams, a game ranger in normal life and an on-call member of the TCU, one day found himself in a sticky situation while on such a listening patrol. Hidden under a village grain bin built between the huts, he had planned to listen to the evening fireside chat and then exfiltrate when all was quiet. This was not to be, however, because, much to his consternation, the entire area's population had been rounded up by a guerrilla group and ordered to attend a *pungwe* (propaganda meeting) in the very village where Russell was hidden.

The meeting, in true African fashion, went on for hours. Not wishing to be found at daybreak, Russell desperately tried to think of a plan to extricate himself out of his delicate situation. About three o'clock in the morning, with the meeting still going full steam, a hush came over the crowd when a senior terrorist leader was introduced. This man, a violent and sadistic murderer, was well known and held in fearful awe by the local people. Mr. Big strutted his stuff, boasting to the crowd how invincible he was and how the soldier's bullets would turn to water if ever aimed at him. With his sweaty face gleaming in the firelight, he regaled the crowd with stories of his prowess at evading security force patrols and of how many sell-outs and policemen he had killed.

As he started to shout political slogans, Russell saw his chance. Abruptly, the leader ceased his spittle-flecked rantings as a 150-grain .308 full-metal-jacketed Winchester bullet entered just under his left eye, blowing the back of his head into the darkness behind him. In the ensuing panic and confusion, Russell managed to escape from his hiding place and disappear into the night.

Russell Williams, a brave and dedicated tracker and soldier, was eventually killed on a follow-up, and this book is, in part, dedicated to his memory.

Search and Rescue

Most jurisdictions that find the need for a tracking team will be situated in a rural district or more remote region of the country. A spin-off job for a tracking team that is of considerable value to any community, especially those with large tourist traffic, is the ability to mount search-and-rescue missions. In all instances where they have been used, trained trackers have proved to be of far greater value than enormous, disorganized search parties in the recovery of lost individuals. Read the excellent book by Jack Kearney, *Tracking: A Blueprint for Learning How*, in which he relates stories of successful search-and-rescue missions conducted by U.S. Border Patrol trackers.

Long-Range Patrolling

Several teams used wisely can patrol large areas of ground and provide good intelligence, thus relieving larger units of routine patrolling tasks. At no other time in the history of law enforcement in the United States has there been such a need for good intelligence about illegal activities going on in remote areas. Marijuana cultivation operations, armed cults, right wing hate groups, and paramilitary organizations have all sprung up like mushrooms in recent years. Mountain states such as Washington, Idaho, Oregon, Montana, and northern California all contain dangerous elements that pose a considerable threat to law and order. One only has to consider the impact of the Randy Weaver incident in Idaho and the Branch Davidian disaster in Texas to be assured of violent incidents occurring somewhere in the United States in the future. Long-range patrolling is an excellent way to gather information in advance to head off other violent confrontations.

Downed Pilot Recovery

Commonplace in wartime, the potential for pilot recovery still exists, particularly in areas where drug interdiction operations are ongoing and aircraft are widely used in spotter and courier missions.

Apprehension and Disruption of Terrorist Infrastructures

With their combination of mobility, speed, and surprise, trackers have been used on a number of occasions to arrest terrorist contact men and supporters, thus removing their logistics and intelligence base. The successful Phoenix Program, conducted by the CIA and Special Forces during the war in Viet Nam, is a case in point where such key figures as commanders, commissars, and tax collectors involved in the local Viet Cong infrastructure were removed, severely disrupting enemy activities.

Investigation of Abnormal Game and Wildlife Patterns

A large group of terrorists secretly infiltrated a remote area of Rhodesia's Zambezi Valley and commenced establishing a chain of bases and arms caches across 60 miles of the valley floor. Due to unseasonable animal movements, a game ranger, also a tracker, was dispatched to investigate. While following strange and suspicious spoor the tracker picked up a piece of cardboard printed with Russian script. Relaying this information back to his HQ by radio, he continued to follow the spoor until he came across a camp where animals, shot by the gang to feed themselves, had been butchered.

Tracking teams were quickly and quietly infiltrated into the valley and soon picked up routes, base camps, and river crossing places without being detected by the guerrillas. Based on the information supplied by the trackers, aerial photographs were taken, pinpointing the terrorist positions and routes. Plotting this information onto maps of the area, intelligence officers were able to site ambushes, pinpoint air strikes, and predict escape routes. At the optimal time, the security forces struck at several occupied bases, forcing the guerrillas into the preselected ambush sites.

Because of the vast and rugged nature of the valley, the operation lasted for almost two weeks. The trackers were prominent in the successful pursuit and annihilation of the entire group of 110 insurgents. Friendly casualties were two killed and seven wounded.

Pseudo ("Sting") Operations

Realizing the need for good and accurate information, Rhodesian military authorities decided to create a secret unit to acquire it. The unit, comprising police intelligence experts, soldiers, and captured guerrillas, used guile and subterfuge to obtain current information on terrorist incursion routes, feeding and staging areas, arms caches, base camps, and support networks. Trackers from the TCU who were fluent in the native languages of Rhodesia and understood the people and customs were assembled and, with the help of a recently captured terrorist to give credibility to the pseudo team, infiltrated the guerrilla infrastructure. So much valuable information began to pour into the intelligence system that special operations were planned to counter the growing threat to the state. (These sting operations eventually proved so successful that it was reliably estimated that over 65 percent of insurgent casualties were sustained in this way, but that is another story.)

Eventually the small band of trackers turned terrorists grew into a formidable unit of close to 1,000. Known as the Selous Scouts after the famous African hunter Frederick Courtney Selous, they were credited with almost 70 percent of all terrorist kills and captures by the end of the war in 1980.

Even though police and corrections SRTs will never be involved in some of these types of operations, it does give an insight into the extreme flexibility and versatility of tracking teams if creativity and forward-thinking policies are adopted.

Training Team Tasks

Rhodesian trackers were often used to train soldiers and police officers from other countries, and this was always a popular assignment. With their bushcraft skills and shooting abilities, trackers were able to pass on their knowledge and tactics to many men. In doing so, they ensured a small but regular flow of high-caliber recruits for the army and their own unit. Irrespective of how long the instruction period was, recipients of this type of training always proved to be better soldiers or policemen as a result of the exposure to innovative ideas, new tactics, and practical experiences.

Chapter 6
ALTERNATIVE TRACKING AND FOLLOW-UP METHODS

Imagination and foresight play important roles in developing new tactics and techniques, adapting them to combat under unusual conditions to defeat an unusual enemy.

Lt. Col. Albert N. Garland, USA (Ret.),
former editor of *Infantry*

Experience has shown that the most versatile, efficient, and successful tracking method is a four-man team working on foot, because there is practically nowhere they cannot go, be it desert, swamp, savanna, forest, jungle, or mountain range. Under certain conditions and terrain, however, there are other follow-up techniques that have proved to be extremely effective:

- Mounted (horseback) follow-ups
- Aerial tracking
- Mobile (vehicle-mounted) follow-ups
- Canine-human combination tracking

It must be stressed that these alternative methods require specific conditions, and if these conditions are not present the follow-up will not succeed and have to be taken over by foot-borne trackers, if not aborted.

MOUNTED TRACKING

We have all seen old Westerns on TV and at the movies where the hero and his faithful sidekick, while pursuing the baddies, stop and "read" the spoor. Lines like "Twelve horsemen passed this way before noon last Tuesday" have become commonplace in Western folklore. As we know from our history books, there is no doubt that tracking skills did play a significant part in the pacification of hostile Indian tribes and settling of the West. The mainstay and major contributor to western expansion was the horse, which provided mobility, endurance, and speed to the cavalry as well as an excellent and stable observation, shooting, and tracking platform.

During the Rhodesian bush war, the need for mounted infantry was recognized due to the vast expanse of the operational areas, and a new mounted unit, the Grays Scouts, was created. Over time the unit reached a strength of 1,000 men. Height and weight of the riders was a major limiting factor in the selection process that mandated a maximum of 180 lbs. for operational troops. This was due to the restricted availability of suitable horses, which, in the main, were smallish crossbreeds good for service in the arid and hostile border areas of the country. These hardy and sure-footed animals were used extensively in the patrolling, recce, and follow-up roles, although they had to be salted, or inoculated, to protect them from the voracious tsetse fly, which could infect untreated horses with "sleeping sickness," the deadly scourge of Africa.

The Grays Scout's mounted patrols became adept at tracking guerrilla groups and racked up a considerable score in enemy kills and captures with very few losses to themselves. Imagine, if you will, being a nationalist guerrilla, bearing a large pack containing land mines plus your rifle, magazines, and ammo, secure in the knowledge that no security forces had been reported in the area, when all of a sudden a dozen or so mounted riders burst out of the bush with weapons blazing. With no place to hide and unable to outrun a galloping horse, your situation would be bleak indeed!

The advantages of mounted tracking are the following:

- Increased visibility for the tracker due to his elevated position.
- Fatigue-free travel—up to 40 miles a day.
- Faster movement, thus covering more ground.
- Dramatic psychological effect on fugitives.
- Ability to carry additional equipment, often up to 300 lbs.
- Self-sustaining up to 10 days (if supplemental food cubes are carried and surface water is plentiful).
- Speed of assault and immediate action drills.
- Horses' good hearing qualities, especially at night.

The disadvantages of mounted tracking are these:

- Dependence solely on available surface water (streams, ponds, pans, or springs) or resupply.
- Expensive to resupply by air.
- Horses' requiring extensive training to overcome reactions to gunfire and battle noises.
- Noisy movement (horses are not tactically conscious).
- Exposed riders, large target.
- No night security possible due to horse noises.
- Horses' natural fear of wild animals, especially mountain lions (although this can be overcome with extensive training).
- Requirement for extensive veterinary facilities and blacksmith backup.
- Horses' fear of fire.
- Requirement for special transport vehicles for uplift to distant operational areas.
- Inability to traverse very thick vegetation and jungle.
- Susceptibility to flies and biting insects in tropical areas.

In an attempt to minimize the disadvantages of the height and size of horse plus rider, the Rhodesians camouflaged their horses with green and brown dyes. As a result of this precaution very few animals became casualties in firefights.

The most critical factor affecting a mounted follow-up is the availability of water. Because many of the areas selected for mounted operations in Rhodesia had sufficient water, the problem was minimal, but horses can consume up to 20 gallons a day, even more if the temperature exceeds 100 degrees. This also means that horses cease to be a viable option in areas where water sources are scarce or nonexistent.

AERIAL TRACKING

Opportunities for aerial tracking (i.e., by fixed-wing or rotary-wing aircraft) are very limited, but since it has been used successfully on occasions, albeit under specific conditions, a few examples are in order to explain how and why this is possible.

Fixed-Wing Aircraft

During the war in Rhodesia, black nationalist guerrillas would infiltrate across the rugged 700-mile border from Marxist Mozambique with the intention of conducting guerrilla warfare against the Rhodesian government and its people. The border between Rhodesia and Mozambique was demarcated by a *cordon sanitaire* consisting of a dirt road running along a cleared 300-foot-wide antipersonnel minefield protected by an eight-foot-high chain-link fence on either side. Guerrillas attempting to infiltrate the country had no other choice but to risk crossing the minefield, and many fell prey to the mines and directional shrapnel devices, leaving easy pickings for the omnipresent vultures, the garbagemen of Africa. Inevitably, some groups found ways to cross the minefield without incident or injury, particularly where wash-aways had occurred.

On one chilly winter morning, an air force pilot flying a Cessna 310 recce plane spotted a minefield breach. A lane had been dug across the field with shovels, and the mines had been removed and placed along the side of the narrow, cleared path. Circling the site, the pilot noticed that when he flew in a northerly direction he could clearly see a line in the tall, dry grass caused by the passage of a number of humans walking in single file. The grass had been crushed down to ground level, and the path was made visible to the pilot by the shadow cast by the standing grass on the eastern side of the trail. There was a number of meandering game trails in the vicinity, but they lacked the regularity and sense of purpose typical of humans on a mission.

Increasing altitude, the pilot circled above and observed more of the path clearly standing out as a line heading toward the Zambezi River escarpment some 20 miles to the south. A quick radio message to his HQ mobilized the on-call helicopter-borne fireforce, which, by good fortune that day, had a tracking team attached. With the fireforce in transit to the area, the pilot continued to follow the trail along the valley floor toward the rugged foothills of the 2,000-foot escarpment and the populated tribal lands on the plateau above.

By the time the troop-carrying helicopters had arrived, the pilot had already computed the grid reference where the path entered the trees and rocks at the base of the escarpment, and the trackers with an eight-man infantry support team landed at that spot. The time was a little before seven o'clock.

Immediately picking up the tracks, the fresh, well-rested tracking team followed the spoor through the morning without a pause along a winding path up the rugged escarpment. A little after 11:30 the trackers realized from the condition of the freshly trodden grass and plants that they were close to their quarry. Opening out into an extended line formation, the patrol crept forward carefully and were rewarded with the sight of the unsuspecting terrorist group resting up after their exhausting overnight dash across the valley floor.

In the ensuing firefight, 29 guerrillas were killed and two captured/wounded. Thirty-two well-filled packs, 23 AK-47s, four RPD light machine guns, 32 landmines, 40,000 rounds of ammo, sixteen 60mm mortar bombs, and two RPG7 rocket launchers with 12 rockets were eventually recovered after a sweep of the contact area. Security force casualties were one noncommissioned officer slightly wounded.

On the face of it, this was a routine follow-up conducted in the traditional way, but the significant feature in this case was the improvisation and skill of the pilot in first locating the spoor and second utilizing the sun angle and shadow in the long grass to track the gang across the valley floor. Had the aircraft arrived only half an hour earlier or later, the sun angle would have been either too high or too low for the tracks to be visible to the pilot from his perch in the air.

Eventually the minefield breach would have been discovered by the routine fence patrol, reported to HQ later in the day, and a tracking team deployed. On this occasion the team was spared a long 20-mile follow-up across the valley floor in extreme temperatures by a switched-on pilot who deservedly shared in the after-action libations.

Aerial Reconnaissance

The police force of Rhodesia was called the British South African Police (BSAP) not because it was British or even South African but because it evolved from a commercial company called the British-South Africa (BSA) Company formed by Cecil John Rhodes to administer the colony of Southern Rhodesia. (Rhodes Scholarships are a legacy of Rhodes and the old BSA Company.) The BSAP, which was made up of both black and white Rhodesians plus a large proportion of Englishman and even several Americans, was an exceedingly competent police force that was well respected by all of the races and tribes throughout the country. It had consistently managed to solve 85 percent of crimes within seven days and was regarded as one of the best police forces in the world. Traditionally the BSAP was Rhodesia's first line of defense, with the army fulfilling the role of "military aid to the civil power."

At the commencement of the bush war in 1966, the police commissioner, anticipating the need for additional specialized manpower, created two new units out of the existing police reserve structure. These were the Police Anti-Terrorist Unit (PATU) and the Police Reserve Air Wing (PRAW). Practically all the pilots of the air wing were either farmers or businessmen who owned small airplanes that were normally hangared in outlying farms and ranches and at small rural airports. At a single stroke, the BSAP created an instant air force with literally hundreds of planes and pilots spread out over the entire country and available on instant call-up. PRAW planes and pilots were used on a rotational basis for courier, recce, airborne relay, and passenger tasks. Some pilots even fitted Browning machine guns onto their planes, thus transforming them into field-expedient gunships!

PRAW pilots residing in the operational areas were often tasked with air recce in the immediate vicinity of their farms and over time developed an intimate knowledge of their assigned zones. Using aerial photographs, they were specifically tasked with looking for new paths and track patterns that could indicate the presence of a guerrilla group in the area. So competent did these pilots and their observers become that on a number of occasions they were able to lead helicopter-borne fireforces right onto occupied guerrilla bases.

On one such occasion in which I was involved, a pilot flying home after several weeks' leave noticed a new path leading from an adjacent tribal land onto a remote, unused area of his farm. Landing at home, he consulted aerial photographs that had been taken several months earlier and saw that no previous track pattern existed in that area, indicating that it was new and most certainly suspicious. He called the local fireforce, which at that time was commanded by yours truly, and we set off to investigate his report.

Circling the area at 800 feet for several minutes I could see the paths but saw no evidence of recent occupation. Rather than write the deployment off as a lemon and it being a slow day anyway, I decided to put a four-man tracking team down to investigate. As the French-built Alouette III descended into a clearing close to the suspect camp, the pilot of my helicopter circling at 800 feet noticed two men wearing farm-type overalls running across a grassy clearing about 600 feet away. One of the fleeing men, looking up at us as we flew overhead, tripped over a tuft of grass and fell, dropping an AK-47 in the process. The game was on!

My gunner opened fire with his 20mm cannon, and I immediately instructed the troop carriers to drop their troops on the ground to form into a sweep line and move toward the suspect camp area. As the troops were deploying, the gunner on my chopper turned his 20mm cannon on two other armed men, both of whom were speedily dealt with. The sweep line continued to move toward the area from where the guerrillas had run and in short order a number of skirmishes developed with several hidden terrorists.

Utilizing well-rehearsed and aggressive drills the troops eventually flushed out and disposed of the gang. In all, a total of nine out of the original 10 were accounted for. Not a momentous engagement, but successful in that the aerial tracker concept had proved itself once again. (We were unaware at the time, but our C.O. had said at that morning's operational conference that he would buy a crate of beer for every guerrilla accounted for on that day. Imagine our surprise and delight when nine crates of cold beer were delivered to the fireforce base that evening!)

Rotary-Wing Tracking

On several occasions my tracking team was deployed to the scene of a contact between terrorists and security forces and, while circling the area, was able to see terrorist tracks. In most cases the guerrillas had run from the scene and the spoor was obvious from the air. Rather than waste time landing, we used the chopper as an aerial platform and followed the spoor for several miles until we had to land, either because the tracks were no longer visible from the chopper or because we did not want to place the aircraft in jeopardy from ground fire. This technique, although valuable in saving time, is not advised against armed quarry unless a considerable amount of time has elapsed and there is little danger of being fired upon by the group: a slow-moving, low-level helicopter is vulnerable to ground fire. If, however, the fugitive group is not armed, the tracks are visible from the air, and frequent confirmable evidence is sought to see if you are still on the correct set of tracks, this is an excellent way for closing the time/distance gap.

Helicopter Leapfrogging

Another useful technique devised by Rhodesian trackers was to use a helicopter to leapfrog tracking teams forward to cut down the time/distance gap when following the trail of guerrillas in uninhabited country. This technique is useful when the tracks are several days old and therefore likely to have covered long distances.

It worked like this. When guerrilla tracks were discovered, a tracking team was brought in to assess the age and direction of travel taken by the gang. After following the tracks for several miles to confirm the direction of movement, the commander would consult his map and make several intelligent guesses as to where the guerrillas may be heading and the route they would have to take to get there.

Plotting this information on the map, the commander would task a tracking team to fly out and search for the tracks along the route, particularly at places like water sources, river crossings, and gaps through hills. Each of these places was scoured for spoor by the team. If nothing was found they would move on to the next point until each was checked out. The tracking team following the original tracks would continue on with the follow-up, first to glean more information about the gang and second to find out whether the gang was lying up in the area. If the heliborne team discovered the tracks at one of the places searched, it would continue with the follow-up while the helicopter flew back to pick up the original tracking team. Repeating the exercise again and again, huge distances were covered in a short period of time and nearly always ended with a successful contact with the guerrillas who were tired from carrying heavy packs on their long march.

On one occasion, tracking teams covered a distance of close to 110 miles through rugged bush country in only six hours, a distance that took the gang four days to walk. This was an important operation, as the gang was in possession of two Russian SAM7 ground-to-air missiles that had been parachuted into Rhodesia from neighboring Zambia. Fortunately, both missiles were recovered in the ensuing contact.

Advantages of Aerial Tracking
- Vast distances can be covered when conditions are ideal.
- The fugitive group is not aware of aerial tracking capabilities and are careless at leaving a trail.
- Helicopters provide a good mobile tracking platform with excellent visibility and communications.
- Use of aircraft has a profound psychological effect on the target group.

- Aircraft, particularly rotary-wing types, are multiuse and can change roles quickly.
- Aircraft can be used to pin down fugitives prior to the arrival of ground forces.
- Aircraft can be used as an airborne radio relay in remote areas or unsatisfactory weather conditions.
- Tracking teams remain fresh and rested until needed on the ground.

Disadvantages of Aerial Tracking

- Fixed-wing tracking is only possible in the early morning or late evening.
- It requires skilled pilots, who are not always readily available.
- It is only possible in country with long grass and limited or no tree cover.
- Circling aircraft alert fugitives to the possibility of a follow-up, initiating countertracking tactics or devices (ambushes or booby traps).
- Aircraft are vulnerable to ground fire.
- Aircraft may be in limited supply.
- Loiter time is limited by fuel capacity.
- Other missions may have priority.

MOBILE TRACKING

The southern African country of Namibia, formerly South West Africa, proved to be a classic operational area for trackers due to the fact that almost the entire country, with the exception of the eastern areas, consisted of flat, sandy, sparsely treed terrain. This fact was utilized to the maximum by the counterinsurgency units of the South West Africa Police, (SWAPOL). These highly trained SWAPOL counterinsurgency teams—known as Koevoet (pronounced "koo-foot," with the koo as in cook), which translates literally from the Afrikaans language as "crowbar"—were the scourge of the Communist-trained guerrillas of the South West African Peoples Organization (SWAPO). SWAPO, through its military wing, the People's Liberation Army of Namibia (PLAN), was attempting to wrest control of the country from the democratically elected multiracial government in a war that was to last more than 25 years.

The operational areas of South West Africa (which borders Botswana, Angola, South Africa, Zambia, and Zimbabwe) were almost entirely desert in the west, changing to grassland interspersed with water-filled pans in the central regions, to the light forests and vast swamps of the eastern regions adjacent to the Okavango Delta in Botswana. Generally, the ground was completely flat and made up of soft white sand. The population was small, a little over 1 million in a country exactly twice the size of California (which has a population of over 30 million). SWAPO guerrillas would periodically infiltrate the country from either Angola or Zambia and walk long distances through uninhabited country to the populated areas, where they attempted to mobilize the tribal people.

As part of the counterguerrilla strategy, the entire northern border of the country was demarcated with a cleared strip of ground 600 feet wide. Police and army groups would routinely patrol the strip by vehicle looking for telltale tracks crossing the soft, sandy strip. When tracks of infiltrators were found, an immediate follow-up would commence to prevent the gangs from reaching indigenous population centers some 60 to 100 kilometers to the south.

Initially using trackers on foot, Koevoet quickly realized that the flat terrain facilitated the use of cross-country vehicles to carry the trackers and support them with firepower once contact was made with the terrorists. After proving the viability of this concept with several successful follow-ups during which a large number of guerrillas were killed or captured, the authorities provided funds for the development of suitable armored vehicles. Requirements included supreme cross-country mobility, speed, and, most important, as near silent operation as modern technology allowed.

In short order the first test vehicles were delivered for evaluation. Known as Caspirs, these armored, mine-protected, passenger-friendly vehicles proved to be outstanding in rapid, almost silent cross-country performance.

With their 20 gears, high clearance, and heavy-duty tires, there was just nowhere they couldn't go. Trees, bushes, streambeds, and broken country were no obstacle to these powerful, versatile, locally designed and manufactured vehicles. Carrying a complement of 14 men consisting of a two-man crew—the driver and commander/gunner—and 12 native trackers, each vehicle was a self-sustained unit. Four vehicles, with a total of 54 policemen, formed an independent "troop."

On operations the troop would prowl along the border until tracks were discovered, whereupon the first 12 trackers would exit their vehicle and commence the follow-up. Spreading out into about a 100-foot line, the trackers would run on the spoor with the Caspirs several hundred yards to the rear, the commanders keeping the trackers in visual contact due to the height of the vehicle.

With the center tracker on the trail, the follow-up would continue unabated. Should the spoor be lost temporarily, the entire line would scan the ground until one of them, finding the tracks, would whistle to his team mates and the formation would move off again at a run. When the first tracking team tired, usually after about 30 minutes, the second team would take its place and so on until contact with the guerrilla group occurred. In this fashion, Koevoet teams could cover incredible distances, often up to 65 miles a day. The speed and efficiency of the follow-up virtually ensured a successful contact with the guerrillas because there was no way they could outrun or outshoot the pursuing policemen.

Once contact was made, the Caspirs would move forward, outflank and assault the guerrillas with their 20mm cannon or .30- and .50-caliber machine guns. The speed of the follow-up combined with the awesome firepower of the Caspirs had a devastating effect on the guerrillas, who tried every trick in the book to attempt to evade or counter their relentless pursuers.

On one follow-up that took place after a small group of guerrillas attacked a military post, trackers covered a distance of about 60 miles in 11 hours. In one week in 1989, Koevoet killed between 1,200 and 1,500 PLAN insurgents who had been sent across the border in an attempt to influence the outcome of the United Nations-monitored election leading to the creation of the multiracial state of Namibia. (For further reading of this little-known but fascinating campaign, read *Beneath the Visiting Moon* by Jim Hooper, an American journalist and the only one ever permitted to accompany Koevoet on operations.)

As can be clearly seen, the mobile techniques devised by SWAPOL can be used in specific terrain conditions. The Koevoet concept was only successful because the ground was suitable for the use of their large four-wheel-drive vehicles. Had there been rivers, rocky outcrops, ravines, swamps, or dense vegetation as existed in Rhodesia for example, the Caspirs would have been only marginally effective, leaving the tracking task to slower but more versatile conventional tracking teams.

The only areas in the continental U.S. where this type of tracking method is viable is probably in the southwestern desert states of Utah, Nevada, Arizona, New Mexico, Southern California, and maybe West Texas.

THE USE OF DOGS IN TACTICAL TRACKING OPERATIONS

This book was originally written specifically about human trackers, but experience gained on the first tactical tracking course ever held in the United States in May 1994 identified and underscored the value of combining human and canine skills to create a total tracking capability. I have tracked on a number of occasions with dogs and have found them to be superb at what they do. However, dogs have some major drawbacks that can and do limit their effectiveness, particularly when used on long follow-ups.

Dogs have the ability to smell up to 10,000 times better than a human being. They are able to follow a trail by sniffing out the residual scent left behind when a person travels over

Scout dogs have played a successful role in following and locating fleeing or hidden fugitives as well as finding bodies, drug and weapon caches, landmines, and ambushes. There is no doubt as to their ability to work well on short-duration follow-ups and for finding the initial direction of flight, but experience has shown that dogs are less effective on longer trails.

the ground. However, they work best in cool, moist, or temperate climates, and should the weather be very hot or windy, no matter how fresh the trail, a dog can lose efficiency quickly. A dog also tends to tire easily in extreme heat and is affected by dusty conditions and wind, and it requires a large amount of water, which has to be carried by its handler. Once a tracker dog loses its ability to track it becomes a liability on a follow-up.

To give a dog its due, it is excellent for short-distance follow-ups and searching for suspects in built-up areas, and it is a superb early warning system. But when it comes to a rural or wilderness follow-up, I prefer to allocate transport space to a thinking, armed, reacting human being. Not that dogs should be neglected altogether—they can be valuable at picking up the initial direction of flight and getting the human trackers off to a good start.

There is a way that humans and dogs can combine their particular talents into an extremely efficient tracking team. This entails training dog handlers in tracking techniques. Once trained, the dog/handler level of ability goes up 50 percent, with the handler adding his visual tracking skills and intellect to the scent-following skills of the dog. It follows then that if the dog loses the scent, there is an excellent chance that the human tracker can continue to follow the trail visually. When the trail enters brush and thick vegetation and the difficulty level increases for the human, the dog can follow the scent. In this way a formidable tracking machine can be developed.

Combining human and canine tracking skills could revolutionize the current law enforcement method of relying solely on dogs at following fugitives. The way things are today, the handler merely accompanies the dog until the fugitive is located; then the handler takes over and makes the collar. This is an area that urgently needs further research and development if law enforcement is keep its edge in the battle against crime.

An interesting sideline to this short discussion of dogs in the tracking role concerns an experiment undertaken by the Rhodesian Army's tracking unit, the Selous Scouts, using English foxhounds. It was suggested by a person familiar with canine abilities that dogs might be trained to track and hunt terrorists in areas of thick bush in the same way they hunt foxes in England. When contact was made, a helicopter-borne fireforce could be deployed to finish off the job. Being innovative and progressive, the C.O. approved the experiment and subsequently 30 imported hounds arrived for training.

Essential to the plan was to have a pack leader fitted with a locating beacon and a voice-activated transmitter so that when the dogs located their prey, they would bark and activate the beacon. The fireforce standing by would be mobilized, home in on the transmission from the beacon, and engage the guerrilla group, who would be, so the plan went, cowering under the assault of 30 yapping dogs.

Although the idea was good (if a little bizarre) in theory, in practice it was a failure. The dogs responded well to training, and a "top dog" emerged to lead the pack, but when tested in the field the whole plan unraveled. Being hunting dogs with many generations of specialized breeding, the hounds, true to their instincts, preferred to follow the scent of deer and small furry animals rather than that of Homo sapiens. There is no doubt, however, that if this experiment had continued over a longer period of time, something more positive would have evolved.

Advantages of Using Dogs in the Follow-up Role
- They are good at short-range trails in optimal conditions.
- They provide excellent early warning capabilities.
- They are capable of finding bodies, wounded people, and equipment caches.
- They cause fear in the pursued.
- They require low maintenance, no pay, no overtime, and no vacations.
- They are fast and aggressive when pursuing fleeing fugitives.
- They are very good at picking up the initial trail and direction of flight.

Disadvantages of Using Dogs
- They tire quickly in hot, dry conditions.
- They need a good water supply.
- They are expensive to purchase and train.
- They are restricted to working with one handler.
- They require veterinary support.
- They are distracted by wild animals and other interesting scents.
- Their handler has to carry food and water in addition to his own.

A Rhodesian police tracker dog boarding a helicopter to accompany a human tracking team to the scene of a contact between armed insurgents and a police patrol.

- They become a liability when tired or distracted.
- They take up valuable space in aircraft or vehicles.
- They are only as good as their handler's physical condition and abilities.

There is no doubt that dogs have a significant role to play in tactical tracking operations, but they are limited to short-duration, highly intensive situations. When it is determined that follow-ups will be lengthy in terms of time and distance, dogs should not be used and the trail left to human trackers.

Chapter 7
COUNTERING THE TRACKER

If you know the enemy and yourself, you will prevail every time.

Sun Tzu

It is obvious that a fugitive from justice who is being pursued by either a police, corrections, or military tracking team will do just about everything in his power to throw off or evade his pursuers once he knows or suspects he is being followed. Depending on the level of desperation to avoid capture, fugitives have been known to try to outrun the trackers, conceal the spoor, lay a false trail, and even wound or kill the trackers.

There are four ways a fugitive can try to throw off a tracking team:

1. *Speed and distance (the time/distance gap).* It is a common belief that for the quarry to move quickly and put distance between himself and his pursuers is an effective method of escape. This is not so, however, because first, speed creates more obvious spoor, which is easier to follow, and second, by the use of the correct tactics, a time/distance advantage can be reduced very quickly.
2. *Antitracking.* This is a deliberate attempt to conceal or disguise the spoor so that pursuing trackers become confused or lose interest in the follow-up.
3. *Spoor reduction techniques.* This encompasses attempts to confuse, overwork, and demoralize trackers. An example is when fugitive gangs, knowing that they are being followed, systematically reduce their number at regular intervals by branching off one or two people until there are no more tracks left to follow.
4. *Countertracking.* These are methods devised to slow down or stop trackers from following up by causing fear and trepidation among the team. Booby traps, antipersonnel mines, and ambushes are all examples of countertracking.

Whatever technique or combination of techniques is used by fleeing fugitives, each creates several advantages, which, if correctly identified, will benefit the pursuing trackers:

- Any form of counter- or antitracking used by fugitives will slow them down and enable trackers to further reduce the time/distance gap.
- All counter- or antitracking methods used

will educate a competent tracker, enabling him to counter those methods and learn more about the fugitive's abilities.
- Irrespective of any method of counter or antitracking used, by employing the correct procedures a competent and aggressive tracker *will* eventually regain the tracks, continue the follow-up, and close with his quarry.

Once a tracker realizes that his quarry is attempting to throw off the follow-up team, it indicates that the fugitive knows or suspects he is being followed. This is very much to the advantage of the tracker in that pressure is placed on the fugitive. Pressure causes a breakdown in the thinking process, compelling people to do things they would not normally do. Fugitives will either attempt to increase speed and therefore leave better spoor indications and deeper tracks, panic and become careless, or waste valuable time attempting to antitrack.

ANTITRACKING

Depending on the ground, vegetation, climate, human activities, and customs in the part of the world where you may be operating, there are many ways that a fugitive can employ antitrack tactics. However, it must be stressed that virtually all antitracking methods can be overcome with the right attitude on the part of the trackers.

The following methods, some good, some bad, have been routinely used in most places where I have conducted tracking operations, and it is likely that similar methods or variations will be found just about anywhere in the world.

Brushing Out the Tracks

Brushing out tracks with some sort of expedient broom made of twigs, leaves, or grass is probably the most common way to try to hide tracks. Its success or failure as an antitracking technique depends on how carefully it is done. However, it will be obvious to the trained eye that the trail has been brushed out by the sudden disappearance of the spoor. Lost spoor procedures will eventually locate it.

I have seen a hunted band of guerrillas drag a large branch behind them in an attempt to brush out their tracks and throw off the trackers. This inefficient technique merely left a very visible trail that was more obvious and thus easier to follow than if they had not bothered to use any antitracking at all. Eventually, the soft pliable ends of the branches were worn down to stumpy twigs, which left clear lines scored into the sandy ground. It was like following a railway line.

Brushing out spoor is wasteful in time as well as energy and is a technique that cannot be kept up for long. The most common use of brushing out spoor is when the tracks cross a dirt road so they would not be obvious to anyone walking along it. The correct lost spoor procedures will eventually relocate the trail under these circumstances.

Once while following a group of 12 guerrillas, we soon became aware that when the spoor passed through native villages the trail had been wiped out and cattle had been driven over the top. It was obvious that the group had instructed the locals to make the spoor disappear. When we encountered this, we merely circled the edge of the village area and quickly relocated the tracks exiting the other side.

Restoring Vegetation

Another common technique is to restore vegetation to its natural state. When a fugitive enters a patch of vegetation, he takes a stick and carefully places the bent and broken stems back into their original position. This is done for several yards until all obvious traces have been obliterated. It is fairly quick to do but will not fool a good tracker for long.

The big problem with this method is that the fugitive is unable to go back to check and see whether he has done a good job because if he does he will leave even more traces. If the fugitive does this again when he exits the patch of vegetation it will be difficult to detect, but remember, he will still have to disguise his ground spoor if he is to succeed in throwing off a good tracking team.

Use of Hard, Stony Ground

This is another common method used nearly everywhere. By crossing over or staying on hard, stony, or rocky areas that do not hold prints, the fugitive believes he can throw off his pursuers. It is probably the most effective method of antitracking, but the success factor depends on the availability and extent of the hard ground. A good tracker should still pick up some clues about the passage of his quarry. Damaged lichen or moss, displaced stones, and scuff marks may be apparent to the trained eye. In the event your quarry selects this type of ground, remember that he eventually will have to leave it. Ground spoor can be relocated by a tracker circling around the hard area. An intelligent appraisal of the fugitive's intentions based on his direction of travel may provide insight into where he may be going, and confirming evidence should be sought there first.

Streams, Puddles, and Waterways

Many old-time Western movies feature scenes of the bad guys riding along a stream or riverbed in an attempt to throw off pursuers. This will or will not work depending on the composition of the streambed and speed of the water flow. Often, if streams are turgid and slow moving, lingering muddy discoloration may reveal that humans have crossed upstream. The presence of crushed and broken water plants and flattened reeds will also show the passage of flat human feet. In some cases the muddy bed of a stream or even a puddle may hold prints for several days. Water splashed up onto rocks may also reveal human passage, and aquatic plants may be dredged up by a careless foot.

The best indicator of the use of a water course to conceal spoor is where the fugitive leaves the water to continue flight on dry land. There will always be evidence left behind at this location, particularly if the banks are steep and scuff marks can be observed. A good clue is mud left on rocks and stones at the point where the quarry exited the water, especially in shady areas or on a cool day when evaporation is slower than usual. Water running out of clothes and footwear may leave signs when it splashes onto the ground. Wet footwear will pick up sand and grit, which will eventually drop off in places where it may be seen. People walking with sodden shoes or boots may stop to dry them out by removing their footwear. This entails a lot of hopping around, leaving a wealth of confirming evidence.

So if fugitives use water or water courses in an attempt to conceal their tracks, remember that they will have to come out somewhere, sometime. By examining both banks up and downstream, the tracks will eventually be relocated.

An interesting case occurred in Malaya during the Chinese Communist insurrection in the 1950s. A platoon of British soldiers, while sitting in ambush at a crossing place on a jungle stream, observed a whitish cloud appear in the clear, fast-flowing water. An examination revealed that the cause was soap from somebody washing not too far upstream from the ambush site. The platoon commander sent a squad of men to investigate.

After working their way through the thick riverine vegetation for several hundred yards, the patrol heard muted human voices ahead. They inched forward to observe a small base camp occupied by a handful of Chinese terrorists, the very people whom they intended to ambush at the river crossing downstream.

Protected by the thick jungle undergrowth, the soldiers carefully positioned themselves on two sides of the clearing. Several minutes later they observed several more armed Chinese arrive. After giving them time to settle down, the patrol threw grenades and sprayed the camp with automatic weapons fire, resulting in the death of seven hard-core guerrillas who had successfully evaded the British for months. Documents recovered from the bodies eventually led to the destruction of the entire gang as well as its support infrastructure. This successful action was only possible because an observant soldier spotted something out of the ordinary and an aggressive commander took advantage of his chance. This is the way guerrilla warfare is won—decisive and aggressive action by junior commanders.

Foot Coverings

On several occasions I have seen attempts to conceal human footprints, particularly in open, sandy areas, either by wearing socks over shoes or by tying strips of cloth, burlap, or animal skins over footwear. This does have the effect of concealing or disguising the sole pattern and blunting any sharp edges, but the addition of extra material on the shoe adds width to the spoor and still leaves recognizable marks on the ground. Marks corresponding to the normal stride of a human are still obvious, and if this method is used in grassy, vegetated areas, clear sign is still left behind.

There is a contemporary British military catalog that features (for $3.50) a set of woven net overshoes designed to disguise tracks. I wonder what genius dreamed that one up. Obviously no one who had ever spent time in the outdoors.

Changing Footwear

This is a common technique in Africa and, though initially quite baffling, is quickly noticeable to the trained eye. Nationalist guerrillas in Rhodesia often carried several sets of clothing and footwear. If they suspected that they were being followed, they would stop and change their outfits. Sometimes they removed their footwear altogether and proceeded barefoot.

If they changed shoes in a place where there was no local inhabitants it presented no problem to a tracker, as the ruse was very apparent. All we had to do was confirm the new patterns and continue with the follow-up. More difficulty was experienced when the fugitives intercepted a well-used path and then changed footwear or went barefoot altogether. It was very difficult to differentiate between local traffic and guerrilla spoor in these situations. However, with a little time spent examining the path and by the process of elimination, it was possible to see where the new prints entered the path. By recording peculiarities of the new patterns such as wear or damage marks, it was sometimes possible to remain on the trail.

Of all the antitracking techniques that I came across on more than 20 years of follow-ups, this was the one that caused more follow-up failures than any other. On one follow-up on which I was engaged, a water truck belonging to the Portuguese Army was destroyed by a Soviet vehicle mine laid on a moonless night on a sandy crossing on the Mukumbura River separating Rhodesia and Mozambique. The local Portuguese commandant sent a runner to my base asking us if we had any spare trackers to assist in a follow-up. Being rather partial to the local *cerveja* (beer) and Portuguese hospitality, I took a full four-man team and crossed the border hotfoot to the scene of the explosion.

We quickly picked up suspicious tracks of one person wearing flat shoes who had entered and exited the immediate area of the mine blast. The tracks led away from the scene downstream along the base of the dry river bank. After following the trail for several hundred yards, it was apparent that the person had laid the tracks at night because obstacles such as bushes, which were obvious in daylight, had been blundered into and not by-passed.

After half a mile or so, the spoor suddenly turned sharp left up though a cleft in the 20-foot-high bank onto a well-trodden path on top. Right on the edge of the path was the clear imprint of large buttocks, more than likely female judging from its size and shape. There was also clear evidence that she removed her shoes there, because a set of fresh barefoot prints headed back in the direction the crime scene. We gained the impression that the person concerned was of a nervous disposition because although the footprints were fairly large, the steps were close and lacked sureness, as if she were afraid of the dark.

Fortunately for us the local population, which was housed in a fenced compound, had not yet been permitted to leave due to the military investigations underway at the mine blast scene. It was easy to follow the spoor back several miles to the outskirts of the dirty, dusty border town of Mukumbura.

A notable feature of this squalid little ville was a brothel and beerhall known as Abu's, which was frequented regularly by the troops of the local Portuguese garrison. As luck would have it, the spoor led us right up to the back door of Abu's place. Having no restrictions to bar our entry, we entered, only to be met by Abu himself and a handful of his "ladies." Knowing from previous experience that attempts to interrogate him would be fruitless, we graciously accepted his offer of a breakfast beer but rejected the offer of his ladies' services. Suitably refreshed by several cans of 3M beer and peri-peri sardines, I walked over to the nearby garrison HQ and reported our suspicions to el commandante himself.

Despite the clear and undeniable evidence before his very eyes, the commandant absolutely refused to believe that one of Abu's whores was the mystery mine layer. After a lot of spittle-punctuated Portuguese cuss words, he eventually shooed us across the dry riverbed back into Rhodesia, concealing under our clothing several liberated Portuguese army ration packs, which were quite a welcome change from our own diet.

I was later informed that the commandant was a silent partner in Mr. Abu's business, which, I guess, was the reason he was reluctant to take any action concerning the identity and origin of his "revenue-generating" mine layer.

Walking Backward

Walking backward is the oldest anti-tracking trick in the book and probably the most stupid. Because of the change of primary impact marks of the sole striking the ground, it is immediately obvious to a tracker when a person is walking backward. A person walking forward places his heel on the ground first, with his weight then revolving from the back of his heel, across the heel, and through the ball of the foot until the toe leaves a final pressure mark as the foot propels the body forward. In the spoor of a person walking backward, the primary impact mark is reversed, with the toe contacting the ground first, thereby pushing sand or soil back in the direction of travel. The pressure then revolves through the ball of the foot to the heel, which leaves a second pressure mark as the foot leaves the ground. Very often there will be a scuff mark where the back of the heel drags soil in the direction of travel.

In addition to the reversal of primary impact marks, a person walking backward tends to take smaller, off-balance steps. A variation on this theme is to tie ones shoes on backward to create the impression of somebody going the other way. Any tracker fooled by this ruse should hand in his badge and hang his head in shame.

Animal Feet

On several occasions in Africa I have tracked fugitives who had tied cattle feet to their footwear to cause trackers to believe that the spoor was laid by livestock. Cattle, of course, have four legs, men have two. If you apply the average pace method to assess the amount of spoor, you are going to come up two legs short. Also, cattle tend to wander aimlessly foraging for food, whereas humans move with a sense of purpose to achieve something or get somewhere.

Elephant and rhino poachers infiltrating Botswana from Namibia in Central Africa regularly used broad, flat elephant soles attached to their feet to avoid detection by government game rangers. For a while this tactic was successful, and their predations of Botswana's valuable game animals continued unabated. It was only when army trackers managed to ambush and kill a poacher wearing elephant feet that the deception was revealed.

The use of animal feet once produced unusual benefits, however. While serving as a company commander in an indigenous South West Africa Territorial Force battalion, I was stationed at a remote bush base a hundred miles from nowhere and close to the international borders with Zambia and Angola. Originally a terrorist-infested hotspot, the area had been quiet for several years, but there was always a potential threat, hence our being there.

I always did my best, difficult with Third World troops, to maintain a semblance of order and discipline, but it became obvious that some bad habits were creeping in. The main one, and to my mind potentially very serious, was the custom of resting one's eyes in the closed position while on guard duty. Despite constant visits by the guard commander and officer of the day, the practice became endemic, and drastic measures were needed if we were not to suffer from a midnight visit by our dangerous adversaries.

I had several sets of cast aluminum animal footprints—lion, leopard, and hyena—which I used when I commanded the Rhodesian Army Tracking School at Lake Kariba. The cast prints were mounted on a frame so that they could be attached on a pair of shoes or boots. One dark night, assured that the guards were busy examining the rear of their eyelids, I strapped on the set of hyena prints and quietly made my way to the perimeter bunkers where the sentries were posted. I left several sets of tracks around each bunker, but at no time was I challenged by the guards, who were, presumably, dozing contentedly inside. The dirty deed done, I went cheerfully to bed.

At the customary dawn PT session, a subdued bunch of soldiers assembled, and it was apparent they were most unhappy. The senior NCO, normally a jovial, happy-go-lucky fellow, approached me and asked if he could talk to me in private. Sending the troops off for a five-mile run before breakfast, I took him aside and asked what the problem was. Gravely, he explained that the local witch doctor, who was losing my soldiers' business to the military hospital, had cast a very bad spell, causing a band of hyenas to come into the base at night to make trouble for the superstitious troops. He took me over to show me the tracks that I had laid the previous night, pointing out that the loathsome animals had confined themselves to roaming around the guard posts.

Taking advantage of the situation, I later told the assembled troops that the reason the hyenas were able to come into the base was because the sentries were sleeping while on duty and that the vile animals had slipped by them in the dark. The black folk of Africa have a great fear and loathing of the hyena, believing the beast provides nocturnal transport for evil witches. Needless to say, we had no more problems with troops sleeping on guard duty.

Smug with the outcome of my ruse, I had my bubble pricked when one of my local Bushman trackers came to talk to me that same evening. Speaking through an interpreter, he politely informed me that he had never seen a two-legged hyena before and proceeded to collapse in a fit of giggling. He, at least, had seen through the ruse.

The point to be made here is that animals, at least the ones you are likely to come into contact with in North America, have four legs. Humans have only two, and if a tracker falls for a trick like that, shame on him! (Did I hear someone mention Bigfoot?)

Custom-Made Footwear

I read somewhere that during the war in Viet Nam, in an attempt to conceal their tracks from the indigenous population, Special Forces soldiers operating in sensitive areas or on cross-border operations were provided with specially made boots with molded rubber soles designed to leave an imprint like a human footprint. Obviously not a lot of thought was used in this project because it did not take long for the locals to catch on to the ruse. The Vietnamese are little people with small feet, whereas the far larger Americans are pretty well-endowed when it comes to the appendage at the end of the leg that comes into contact with the ground. Not a smart idea at all, and typical of ideas emanating from the mind of an armchair commando.

Eventually common sense prevailed. Somebody came up with the expensive idea to air-drop hundreds of pairs of boots into the area with the same pattern soles as those worn by the soldiers. Grateful for the gift from heaven, the locals wore the boots and, in doing so, unknowingly masked the tracks of the Green Beret reconnaissance teams. Frankly, they would have been better off wearing locally manufactured Ho Chi Minh sandals made from old car tires!

Walking Along Paths and Tracks

A common antitracking technique is for fugitives to join a well-used path so local pedestrian traffic will cover and obliterate their spoor. (A variation on this method is to walk along wheel tracks, knowing that later wheeled traffic will cover or dust out the spoor.) This is a successful technique and the one most difficult for trackers to overcome. The only way to really get over it is to carefully examine the path for several hundred yards in either direction in an attempt to isolate and identify the correct prints from those of the local folks. This can be a tedious and frustrating process, and success or failure depends on how many people have used the path behind the fugitives and prior to your arrival.

On one occasion, my team was following a trail that intersected (probably on purpose) a railway line, allowing the group of terrorists to walk along the railroad ties to conceal their spoor. Initially the tracking was easy due to the sand and dust deposited from their footwear, which left a faint but discernible trail on the dark wood, but with time and distance the trail got fainter and eventually disappeared altogether. We never caught up with them, and it was one of the few times that we were forced to concede defeat.

Fugitives can totally conceal their trail effectively by walking along paved roads, making it difficult or impossible to track them. The best a tracker can do under these circumstances is to laboriously search along the roadsides in the hope of finding evidence of exit in the same way a search is done along riverbanks. This can be a time consuming, exhausting process and is tracker intensive. The disadvantage to the tracker is that the time/distance factor increases to the advantage of the fugitives because they can make their escape while the trackers spend valuable time searching along the road. On several occasions I have had to abort follow-ups because the fugitives were able to get a lift on a passing vehicle driven either by a willing or coerced accomplice.

Abrupt Changes of Direction

A simple but effective technique that can fool even the most aggressive tracker, if only briefly, is for the fugitive to abruptly change direction, especially if done on hard, unyielding, or rocky ground. Using the correct lost spoor procedures will eventually regain the trail, but abrupt changes in direction can often delay a good team, resulting in an undesirable increase in the time/distance gap.

The sudden switch of direction is usually by about 120 to 170 degrees, and the quarry moves off on a different tangent. By executing several of these switches, terrorists have been able to delay less well-trained tracking teams, who tend to overshoot the last known spoor. Unless the change of direction is picked up immediately by one of the flank trackers, the tracker will suddenly run out of spoor and waste valuable time looking along likely directions before commencing his 360 search pattern. Eventually he will regain the spoor, but in his heart he'll know he was fooled. Should the fugitive group pull this trick more than once, though, the tracker will become aware and immediately commence his 360 search *back* along the trail instead of forward or request that his flank trackers assist in looking for the breakaway spoor.

Other Methods

On one occasion it was obvious that a small group of guerrillas I was following had clambered along a barbed wire fence during the night for over 1,500 feet, hoping not to leave obvious spoor on the ground. Their efforts, which must have been exceedingly slow and painful, were a waste of time because in several places we found small strands of cloth torn from their clothing left on the barbs. By moving quickly along each side of the fence we soon picked up the trail where they eventually left the fence line. We calculated that the group must have lost about an hour of valuable time performing their time-wasting wire-walking act.

There is a wide variety of antitracking techniques that will work to a certain degree depending on the terrain or the skill of the evader, but a good tracker will eventually learn to overcome each and every obstacle thrown in his path and soon be back on track. Let me stress this again: on all of the courses I have run in the United States, I have had trainees almost give up on a follow-up, claiming that they had terminally lost the spoor. In every single case, after utilizing the correct lost spoor procedures, they relocated the tracks. True, it may have taken several hours, but the point is: *these techniques really work*. It just takes practice and more practice.

SPOOR REDUCTION TECHNIQUES

Spoor reduction techniques are an effective way to delay a follow-up and frustrate a tracking team. Basically, spoor reduction techniques entail the pursued group's simply splitting up into smaller and smaller groups until either all available trackers are on the ground following different sets of spoor or there are no further tracks to follow. There are three specific techniques: bombshelling, breakaway groups, and drop-offs, all of which have been used to confuse and obfuscate trackers with varying degrees of success.

Bombshelling

This is a common spoor reduction method, especially in areas where guerrilla bands are familiar with follow-up techniques employed by security forces. Bombshelling consists of each person in the group leaving the scene in a different direction to eventually meet up at a prearranged rally point.

During the war in Rhodesia, it was army policy that each action, every sighting, or even the faint possibility of a terrorist presence be investigated. This factor made terrorist groups fearful of putting themselves in a position where they could be tracked and possibly annihilated by aggressive helicopter-borne fireforce units. Whether it be laying a landmine, murdering a "sellout," attacking a farm, or ambushing a convoy, terrorists knew that the first security forces on the scene were likely to be trackers and that a follow-up team would soon be on their trail.

Bombshelling made the tracker's job difficult for several reasons. First, he had to correctly identify the spoor to be followed. Second, it was difficult to estimate the number of guerrillas involved in the incident. And third, it was difficult to estimate their direction of flight. With a group of 10 to 12 terrorists, which was the norm, trackers would be forced to take more time than normal obtaining critical elements of information than would have been the case if the terrorists left the scene in a single group.

After the trackers had obtained the best guesstimate of the number of guerrillas involved, they then had to decide which set of prints to follow. Usually they would choose the most clear and obvious sole pattern made by a large, heavy individual with big feet. Even then it was not possible to confirm positively whether this set of tracks was made by one of the terrorists and not by an innocent local who had fearfully fled the scene.

Often under these circumstances, interfering HQs would often order the team to split up and follow different sets of spoor in several directions assuming that they would eventually meet up somewhere. It is important that trackers be left to make their own decisions under these circumstances, bearing in mind that it is not wise to break up the integrity of a close-knit and well-trained tracking team. It is far more efficient and therefore quicker to put all of the available tracking resources onto one spoor and attempt to move quickly to the terrorist rally point than risking the lives of the team by requiring them to track alone.

Breakaway Groups

This technique is basically the same as bombshelling but consists of small groups rather than individuals breaking away from the main group. When only one team of four trackers is involved on the follow-up, it causes the problem of which group to follow and which to ignore. We have already discussed the inadvisability of breaking up a tracking team when faced with multiple trails, so it is essential under these circumstances that the whole team confer and decide on which set of tracks it should follow based on the best chance of success. If other standby tracking teams are available, the possibility of bringing them in should be considered. However, this always creates a dilemma that has to be resolved quickly because it is entirely uneconomic to use scarce tracking resources that may be better employed elsewhere. If a decision has to be made, it has to be made by the team members on the ground, who will have a better feel for the tactical situation, and not by some distant, out-of-touch HQ.

As a general rule in breakaway situations, it is best to follow the largest group, but you must always bear in mind the possibility that this group may eventually break up again into even smaller groups. The same principle still applies—follow the trail that you think will give the best chance of success. Should a guerrilla group use breakaway tactics in an attempt to evade trackers, it indicates that it has a good knowledge of your follow-up tactics and tracking abilities.

Drop-Offs

The drop-off method can best be described by the following story. I received a radio message that a police base had been attacked by a large terrorist group using fairly accurate mortar fire, resulting in several officers killed and a number wounded. Tasked with the follow-up mission, my team was uplifted by helicopter at first light on what was a cold and frosty morning. Even in Africa the weather can get bitterly cold in the winter months, and on this occasion my lightly clad trackers, wearing only shorts and camouflaged t-shirts, shivered in the icy windchill as the doorless chopper flew over frost-encrusted ground at over 100 knots per hour. I knew that we had to put in a maximum amount of effort on this follow-up because several police officers had been killed, emotions were running high, and, more important, the attacking terrorists had used Soviet-manufactured 80mm mortars, which was the first time we had encountered these accurate and powerful weapons. It was important that we attempt to capture the mortars and take them out of circulation.

While circling above the base we could see a large number of mortar splashes forming a line between the sandbagged perimeter and a small rocky outcrop some 500 yards away. It was obvious from the air that the terrorists had lined up their mortar tubes on the lights of the base and walked the shells right into the perimeter. Fortunately, most of the splashes were outside the base perimeter, but the ones that had hit inside had done a large amount of damage in the lightly protected tented accommodation area.

Gulping down hot but welcome coffee, I was briefed by the base commander about the circumstances of the attack, and shortly after we set out to investigate the rocky hill that we had observed on our incoming flight. Examination of the hill confirmed that one 60mm and two 80mm mortar tubes had been used in the attack and about 20 guerrillas had been involved. Backed up by a squad of angry and vengeful police officers, we commenced to follow the trail, which was easier than usual due to optimal tracking conditions. The low angle of the early morning sun revealed clearly the deep prints of heavily laden men: 80mm mortars, complete with base plates and bipods, are heavy pieces of equipment to carry around.

The spoor headed off toward a range of brooding, tree-clad hills some two miles distant. Making good progress, we quickly reached the base of the hills, still enveloped in early morning shadows. As the trail headed up the rocky slope, the ground conditions changed abruptly and the spoor became difficult to see because of the multitude of shadows creating a confusing dappled effect on the ground. Changing tactics, I brought the flank trackers in and we made better progress by covering a wider front.

As the follow-up commenced into the morning we slowly became conscious of the fact that the spoor was becoming less and less distinct. After discussing this phenomena, the team came to the conclusion that at about every 800 to 1,000 yards, one of the fugitives was dropping off the trail and antitracking away from the main group by using rocks and the hard sun-baked ground. This assessment was finally confirmed when, having no alternative other than to stay with the largest amount of spoor, we ended up following the tracks of just one person who eventually joined a well-used path, removed his shoes, and joined the throng of peasants on their way to a local administration center. It was obvious that the group did not want to risk losing their new mortars, and by using unprecedented anti-tracking skills, they managed to outwit us poor trackers. The only alternative left to us was to return to an earlier section of the trail, seek a specific set of prints, and attempt to follow them to wherever they may lead.

The drop-off technique is an effective method of evading trackers, especially if used in rocky terrain where it is easy to break off from the group leaving little or no sign. One effective counter is to return to an earlier part of the trail, but doing this puts the time/distance initiative back in the hands of the fugitives. Another way that has been successful on several occasions is to backtrack along the original trail used by the gang. On the downside the tracks will be older, but it is likely that no attempt would have been made to antitrack, so the follow-up will be quicker. Backtracking along a trail has often revealed a lot of clues about the gang's origin and provided invaluable intelligence about where it makes its base and feeds. This information, when processed into intelligence, has often resulted in operations leading to a gang's eventual location and subsequent destruction.

COUNTERTRACKING

Countertracking is an offensive tactic used against trackers in an attempt to delay or abort a follow-up by causing injury or death (or the fear of injury or death) to members of a pursuing tracking team.

Should a member of a team be killed or injured by a countertracking technique while on follow-up, the casualty will have to be medevaced out of the area. This will have several benefits for the fugitives:

1. They will have slowed down or even stopped the follow-up.
2. They will be aware of the exact position of the follow-up group either by the sound of the explosion or the descent of the medevac helicopter (assuming one is used).
3. They will have reduced the tracking team strength by the amount of casualties sustained, thereby diminishing its effectiveness and morale.
4. Their action will have a severe psychological effect on the team, perhaps even to the point of provoking anger, which may lead to an unwise emotional response, or fear, which may lead to paralysis or lack of determination.

Veterans of the war in Viet Nam can testify to the damaging psychological effect of randomly placed booby traps and mines killing and maiming indiscriminately. But the tracker, when following spoor, has a much more personal war and must contend with the fact that any countertracking tactics used by the fugitives are directed specifically at him.

The countertracking techniques described in this section have been used extensively by Asian, African, and South American guerrillas trained by either Chinese, Cuban, or Warsaw Pact advisors, so it can be assumed that similar tactics may be encountered all over the Third

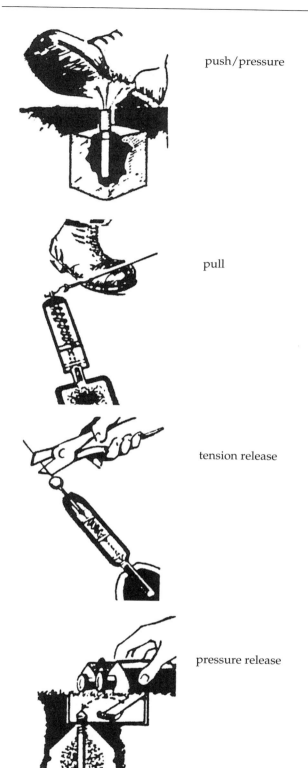

World where Communist military advisors have been spreading their trade secrets.

Although it is considered unlikely that fugitives from justice in the United States would use countertracking tactics against a follow-up team, it must be remembered that marijuana growers in remote areas are making extensive use of booby traps to protect their secret cultivation plots from detection by law enforcement or accidental discovery by hikers or hunters. The possibility definitely exists that such measures will be used against tracking teams operating in grow areas, particularly when following the spoor of growers into or out of their area of operations. The good news is that of all people involved on drug eradication programs, a trained tracker is more likely to spot lethal booby traps than anybody else.

It is also important to note that survivalist and other potentially violent groups are flourishing in the United States, particularly in the northwest states where local political sympathies are supportive of antigovernment attitudes and the terrain is ideal for irregular warfare. Even as this book was being written, there were reports of militia groups, organized to oppose unconstitutional gun control efforts by an increasingly intrusive government, springing up all over the country. These militias are, in the main, completely legal and pose no threat to law and order, but renegade groups and fringe elements can and do have the capacity in terms of manpower, training, and weaponry to take on the forces of law enforcement in what could be a protracted and bloody struggle. To complete this gloomy scenario we must also add the threat of hostile Islamic fundamentalism, which has already

Figure 13: Mechanical booby trap switches of U.S. military origin. There are four basic ways a booby trap can be initiated mechanically: push or pressure, pull, tension release, and pressure release. The U.S. military has devices, known as switches, that accomplish these functions. All are available from surplus stores or can be acquired illegally from servicemen.

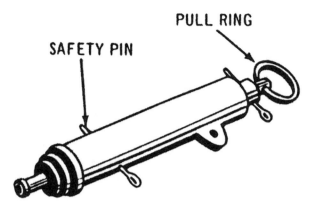
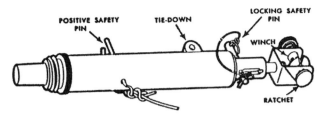
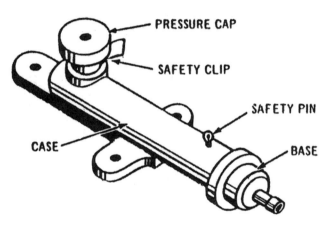
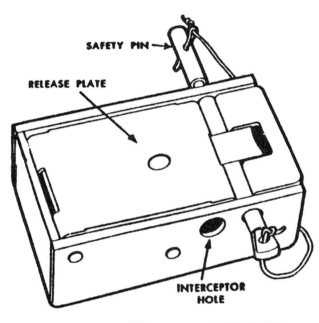

Figure 14: U.S. military booby trap switches.

reared its ugly head in our cities, and the growing threat of street gangs migrating into rural areas and training in our national parks and forests.

Whatever the scenario, we must remember that there *is* a danger that law enforcement officers may have to face armed irregulars in the not too distant future. Let's hope then that police and sheriffs' departments in remote areas have taken the necessary steps to train their officers in counterguerrilla warfare and rural tactical operations.

Antipersonnel Mines and Devices

Antipersonnel (AP) mines come in a variety of designs, from wooden boxes to molded plastic models and even improvised devices using field-expedient materials. An understanding of how these devices can be activated will lead to a better idea of where they can be placed and is the first lesson in any mine-awareness program.

Trackers should *never* attempt to neutralize AP mines or devices by exploding them, as this will give away the team's position to the fugitives. Mines should be marked clearly and quickly so that any supporting element following behind the team does not activate them accidentally. If an antipersonnel device is found, it should be photographed if possible

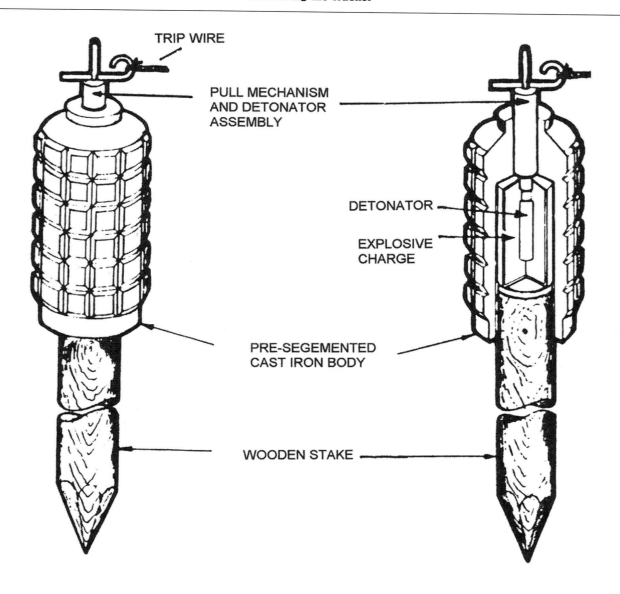

and the information passed on to other departments and units as well as the FBI and BATF. *All* booby trap or mine sites must be recorded accurately on a map and the information passed on to bomb disposal personnel who are trained to handle them. (Switches used in AP devices are shown in Figures 13 and 14.)

Pressure Devices

Direct pressure, usually applied by the weight of the body through the foot, will activate this type of device and lead to

Figure 15: Pom Z shrapnel antipersonnel mine. This AP mine is of Communist bloc origin and can be found on all continents. Known as the "chocolate box mine" (because when it goes off, everybody gets a piece!), it has been used by guerrillas worldwide for many years. A cast-iron body with a removable cylindrical explosive charge is placed onto a special wooden stake that is implanted into the ground. The activating mechanism is either pull or pull/release, making it ideal for use in rural and wilderness areas. Its effective casualty area of about 30 feet results in serious shrapnel wounds to the lower body. Its effect on a tracking team will depend on the formation adopted, so it is likely to be placed in an area where movement is restricted or channeled.

traumatic wounds requiring immediate evacuation. Designed to do severe damage to the lower extremities, pressure devices can be difficult to locate if well concealed and are often laid directly on the trail to maim any pursuer. A bear trap is a typical example of a mechanical pressure device.

Pressure Release Devices

A variation of the pressure device. An example is the German "Bouncing Betty" of World War II, which was activated only when weight of the foot on the pressure plate was released. The mine was then blasted several feet into the air by a small black powder charge, whereupon it exploded.

Pull Devices

These devices use a pull on a trip wire to activate their mechanism. Wires are usually laid across the trail, and an explosive device, which is set alongside the path, detonates close to the activator. Trip wires are often camouflaged but can be detected by trained trackers. A Soviet POM-Z is typical of this type (see Figure 15).

Pull/Release Devices

These are trip wire devices that are activated either by tripping or cutting the wire. No attempt should ever be made to cut any trip wire in case it is of this type. Wires should be followed carefully to both ends and examined closely to ascertain the type of mechanism used. Only then can the wire be cut if it is safe to do so.

Booby Traps and Improvised Explosive Devices

In the vast majority of cases in the United States, the tracker will be following people who have no idea they are being trailed by a skilled follow-up team. It is unlikely that a tracker will ever have to contend with a fugitive with the knowledge and skill to harm or even kill him with a lethal booby trap. This is not the case, however, in areas used for marijuana cultivation or if renegade militias or militant survivalists are involved. The same applies to a military follow-up team engaged in a low-intensity conflict against trained guerrillas. Bearing in mind that a tracker should always expect the unexpected, acquiring information on booby traps is an important part of his training and may make the difference between life and death on operations. Knowledge of these principles will help him anticipate where booby traps are likely to be laid as well as detect them in place.

The Principles of Booby Trapping

There are a multitude of ways that booby traps can be disguised, hidden, and activated (see Figure 16). To become more conscious of booby trapping it is well to have an understanding of the military principles of this lethal art. These principles include the following:

1. *Appearances.* Concealment is mandatory to success. All litter and other evidence of booby trapping must be removed. All this means is that if any litter, soil, evidence of digging, cut twigs or branches, or anything else is out of place, it will attract the attention of the tracker and alert him. This will make him aware of the possibility of a booby trap, and he will take the necessary precautions.

2. *Firing.* An obvious firing assembly may distract attention from a cunningly hidden one. Booby traps often have been placed where they are easily seen. With his attention on the obvious device, a person investigating the booby trap triggers a more cunningly concealed device.

3. *Likely areas.* To be effective, booby traps should be placed where they can be expected to cause casualties. If it is suspected that pursued fugitives may place booby traps, the tracker must be extremely conscious of places where the team may be channeled into a constricted area. This could be between two closely spaced trees, a narrow gully or defile, a streambed, or any other place where the path narrows due to the terrain or vegetation. Other good places to site booby traps include where the spoor enters a dark, shaded area or

1. An electrically activated pull switch utilizing a common clothespin. A pull on the trip wire will remove the interrupter from the contact and complete the circuit.

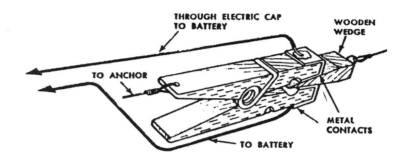

2. A knife blade can be used to create an effective tension release or pull trigger. Whether the trip wire is cut or pulled, the circuit is completed and the device will detonate.

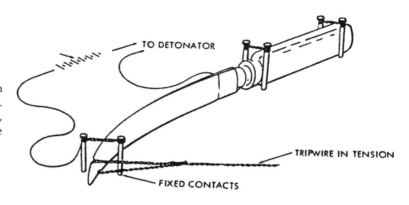

3. A clothespin can also be used as a tension release trigger when coupled with an electric circuit. If the trip wire is cut or broken, the spring closes the pin, thus completing the circuit.

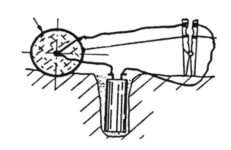

4. A grenade with the pin removed and attached to a trip wire is placed in a can. When the trip wire is tugged the grenade is pulled from the can, allowing the lever to fly free and detonating the grenade.

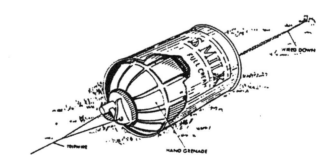

Figure 16: Improvised explosive device triggers. Many examples of improvised explosive devices (IEDs) have been recovered from marijuana cultivation areas. Designed to inflict casualties on law enforcement officers on drug eradication programs and to keep away hikers and hunters, IEDs can inflict serious injuries and even death on the unwary. They employ the same mechanical actions as military switches to activate. Many household and workshop articles can be used to fabricate IED triggers; the examples here are typical of what can be used.

Figure 17: The Claymore antipersonnel directional mine. The Claymore, an effective antipersonnel mine designed to blast a hail of steel balls in a given direction, is lethal to up to 50 yards. Claymores can be initiated by tripwire or detonated on command by either an electric cable or a radio-activated device. Claymores can be fired singly or joined by detonating cord to cover a wider front.

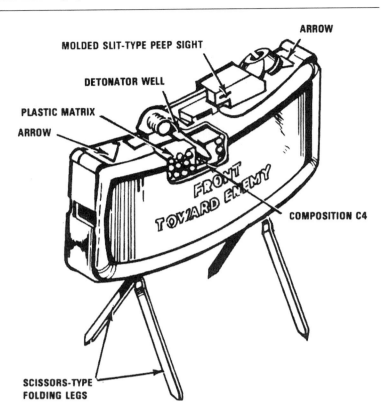

emerges into sunlight where the eyes have to adjust to the different light intensity. This is also why trackers should never walk on top of the fugitive's tracks.

4. *Obstacles.* Road blocks, fallen trees, and litter are all ideal locations for booby traps. When a tracker notices an obstacle ahead, he should always stop the follow-up and inform the controller, who will detail the flank trackers to investigate. As in principle 3 above, obstacles create opportunities to inflict harm on a follow-up team.

5. *Gathering places.* There is no reason why a follow-up should not be conducted in built-up areas, particularly around small rural or wilderness communities. Fugitives will always head for places where they can obtain assistance, shelter, food, and water. Besides buildings, building entrances, and similar places where people gather, this principle also applies to shade-giving trees or rocks on a hot day as well as access points to water sources. These are all places where an intelligent enemy will place deadly devices to catch the unwary.

6. *Appeal to curiosity.* Booby traps laid in bold positions to dare the curious get results. It has been emphatically stated elsewhere in this book that trackers must never touch anything left lying on the trail until the proper precautions have been taken.

7. *Bluff.* Dummy booby traps, repeated consistently, may encourage carelessness. An obvious booby trap may mask another more deadly one. Sometimes called "double bluff," experts at booby trapping will go to any length to harm their enemy.

8. *Lures.* Booby traps may be baited. The unexpected detonation of an explosive device may scatter troops or detour them into a more heavily laid area. The history of mine warfare is full of stories where victims of one mine have sought protection in a place anticipated by the enemy for such a purpose. What seemed to be a safe refuge turned out to be a lethal killing ground!

Distraction techniques used in conjunction with booby traps need to be emphasized here. Fugitives may attempt to delay a follow-up by leaving a booby trapped item, seemingly dropped or discarded, that is likely to attract the attention or curiosity of the tracker. One common method is to place a pressure release device under an attractive piece of gear such as a pack, binoculars, or weapon. When the item is lifted or moved the device explodes. Trackers must not be tempted to handle what appears to be discarded equipment *no matter how attractive or innocent it may seem.* If the item must be examined or if you must have a souvenir, take all the necessary steps to safely

Figure 18: The Claymore is provided with an accessory bag containing a firing device, firing cable, connectors, and stakes. The electric detonator is housed in the cable spool.

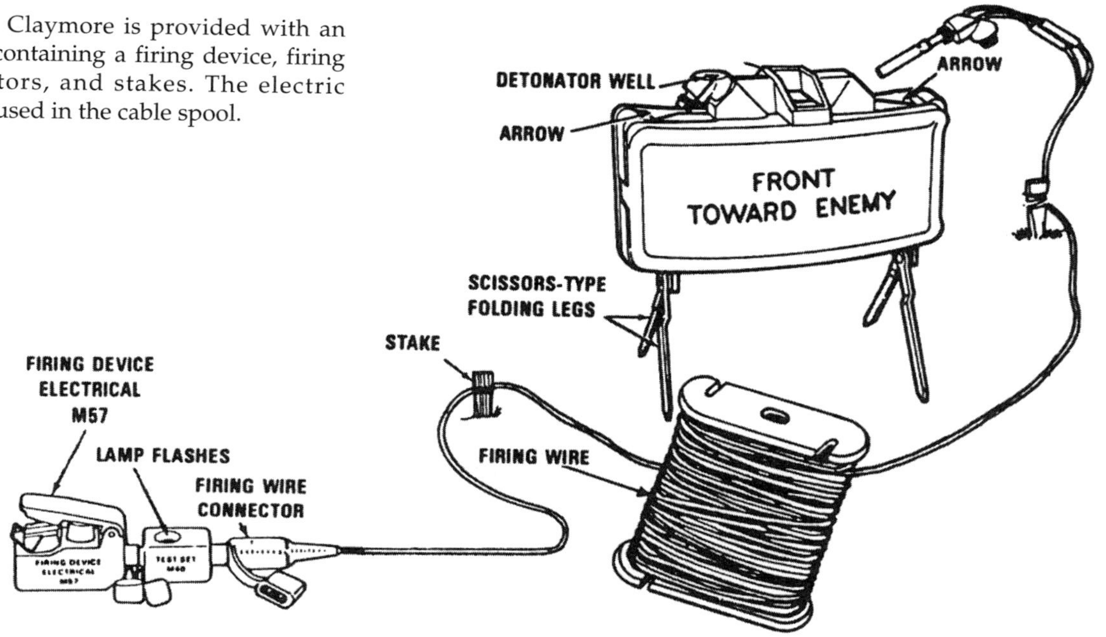

move it: attach a line to it, hide in a safe place, and pull the line. If it is booby trapped you will at least survive the blast. (Unless, of course, the fugitive has correctly anticipated your behavior and placed an explosive charge where you are hiding. It has happened before and will happen again.)

On one occasion a new member of my unit found an AK-47 concealed in a thicket of reeds. Gripping the weapon by the barrel, he attempted to pull it out, whereupon the thing fired. A hot, steel-cored, 125-grain bullet flew at 2,350 feet per second between his arm and body. Had the weapon been set on full auto he may not have survived to suffer the embarrassment of his folly. Treat all discarded equipment with extreme caution until you have thoroughly checked it out.

Again, unless proved otherwise, booby traps are not likely to be a threat in the United States. Complacency kills, however, and trackers should always be alert to the possibility. Devices need not be of the explosive type, trip wires need not only be at foot height, and traps can be disguised as anything. Stay awake, stay alert, and stay alive.

Command-Detonated Devices

Nearly all booby traps and antipersonnel mines are activated by direct mechanical action, usually when trodden on or moved by human activity. As such they are indiscriminate and can be triggered by friend, foe, or animal. Devices of this nature certainly do present a problem for the tracker, but with a good understanding of the principles of siting booby traps and a keen eye, most can be found prior to activation.

Not so is the case of a command-detonated device such as a Claymore mine (see Figures 17 and 18). The Claymore is an antipersonnel shrapnel device designed to blast a mass of lethal, high-velocity steel pellets in a desired direction. Any living thing unfortunate enough to be within the killing area at the time of detonation is not likely to survive.

The term "command detonation" means that the device can be activated on command either electrically or by radio signal when the majority of the target group is in the killing ground. In Viet Nam, Claymores were used with great success by Ranger long-range reconnaissance patrols in the ambush role, and many a Viet Cong guerrilla or North Vietnamese regular came to a sorry end never knowing what hit him.

Just as our soldiers employed the Claymore so effectively, it can be used on unsuspecting trackers too. All the operator has to do is to sit and wait until the trackers enter the killing zone, then detonate the charge.

Claymores are easily concealed off the trail and, if intelligently sited, can be difficult to find. A good flank tracker should be able to pick up the wire or tracks leading away from the ambush site, but by then it is probably too late. If you can see the wire, chances are the operator can see you and will not hesitate to detonate the mine. If the mine is to be detonated by radio signal, there is no way of knowing about it except for the possibility of picking up the tracks of the person who placed the mine entering and leaving the area. Improvised exploding ambushes can also be constructed by linking hand grenades or other explosives together with detonation cord and triggering them the same way as a Claymore.

Acting on good intelligence, a Rhodesian Army special operations team was tasked with ambushing a convoy carrying guerrillas who were expected to travel along a certain dirt road prior to infiltrating across the border from Mozambique into Rhodesia. Hoping to destroy the vehicles as well as the occupants, they buried some 50 lbs. of plastic explosives in the center of the road. An electrical trigger connected to the mine by a thin wire was to be used to detonate the charge. To conceal the wire from casual observation, it was buried under the road and into the brush for several yards. Disguised with vines and grass, it led to a vantage point that gave the team good visibility for several hundred yards in either direction.

After waiting for several days, the team was alerted by the sound of a vehicle engine in the distance, but they were surprised to see a four-man patrol moving along the track some 300 yards ahead of the convoy carefully scanning the ground. Realizing that the patrol would notice the disturbance in the road, the team took up their weapons and prepared to fight it out.

Just as they feared, the foot patrol noticed the disturbed ground and proceeded to investigate the surrounding area. One of the soldiers called excitedly to the others as he discovered the hidden wire and started to scan the high ground above the road in the direction from which the wire originated. Realizing that the game was up, the team leader pressed the button of the detonating device and a massive explosion blasted tons of rock and gravel into the air. When the dust settled, there was not a single trace left of the foot patrol, which was atomized by the blast. The convoy, seeing the explosion ahead, swung around and was last seen heading rapidly in the opposite direction.

Learning from the experience, the unit modified its procedures and started to use a radio-activated detonation system.

In another incident, a 10-man team infiltrated across Lake Kariba into Zambia by canoe and prepared an explosive ambush for an enemy truck expected to pass along a dirt road leading down to the shore of the lake. After lying in wait for close to a week, the team was alerted to the arrival of its target by the sound of a truck engine in the distance.

Experience had taught the team members that it was better to ambush moving vehicles on a downhill grade so that even if the explosives didn't finish off the vehicle, the subsequent crash would add insult to injury. Accordingly, they had buried close to 60 lbs. of plastic explosives in the center of a steeply descending track and planned to detonate the charge by radio signal as the truck bed passed over the bomb.

As the engine noise grew louder, a lookout stationed on top of a rocky outcrop reported that three trucks, not one as expected, were approaching at around 10 miles an hour and no foot escorts were visible. He described the vehicles as large, uncovered, and traveling about 100 feet apart. Unexpectedly, each vehicle was loaded with between 50 to 60 armed men. Realizing that the tactical situation had deteriorated radically out of his favor, the team commander rapidly deployed the remainder of his men into an ambush position to engage the second and third trucks with machine gun fire if necessary.

With hearts pounding, the team watched the first truck come around the bend and descend the slope. As the vehicle entered the killing ground they were amazed to see that apart from being loaded with 55-gallon fuel drums, uniformed guerrillas were seated on top of the drums and squeezed in everywhere they could find a place to sit.

At just the right moment, the vehicle vanished in a fireball. The team, lying in cover some 30 yards away, felt their eyelashes and hair being singed as a wall of scorching flame passed over them and secondary explosions blasted in the killing ground. Fingers on triggers, they awaited the anticipated reaction from the second vehicle, which had come into view around the bend.

To their utter amazement, they watched the driver leap from the cab of the still moving vehicle, which rammed into the flaming wreck of the first truck. Pouring a deadly hail of bullets into the packed truck bed, the ambushers were again assailed by a wall of flame as the second vehicle exploded into a fire ball. No more than 10 seconds had passed and already two trucks were ablaze and over 100 men were dead or severely wounded.

Realizing that there was no threat to its front, the team turned to face the third truck, which, incredibly, was rolling downhill, driverless, towards the flaming wreckage! With its passengers screaming in terror, the third truck crashed into the white-hot wrecks and burst into flames. The guerrillas that were able to leap off the back and flee in panic down the rocky slope were pursued by a stream of .308 bullets, tracer rounds, and rifle grenades.

It was impossible to count the dead, but it was estimated that close to 140 out of the original 160 guerrillas had perished, and three five-ton Scania trucks were destroyed in less than 30 seconds of mayhem. A successful operation . . . but the story does not end there.

As the team moved to an exfiltration point several miles away to be uplifted by a helicopter, a Zambian Army vehicle approached the ambush site from the opposite direction and was able to inform its HQ by radio of the disaster that had just taken place.

As the tired but elated team was flying back over the lake toward Rhodesia and a welcome cold beer, the pilot motioned to the team leader to put on a headset and await a message from his commanding officer back at base. He was informed that a Rhodesian intelligence unit had monitored the Zambian Army communications and learned that the supreme guerrilla commander himself was on his way to examine the scene of the ambush. Welcoming the chance to inflict further damage to the enemy, the team leader instructed the helicopter pilot to return to Zambia, where they would attempt to ambush the leader along the same track.

Landing in the growing darkness some distance away, the team set out on a forced night march back into the ambush area. Low on ammunition and unsure from which direction the commander would be approaching, the team buried several Soviet antivehicle mines on the track to the south of the original ambush site and lay up in ambush on the northern approach.

Around noon the next day, the team heard an explosion to the south, which they assumed was one of their mines detonating. Half an hour later they were informed by HQ that the vehicle (a 10-seater Land Rover) had in fact struck a mine and that *nine* senior guerrilla leaders riding in the vehicle as well as the top man himself were either dead or seriously injured. Elated with success for the second time in 24 hours, the team was again uplifted by the same pilot as the day before and set out for Rhodesia.

Halfway across the lake, the pilot motioned to the team leader to listen in on the radio for another message from HQ. With a sigh of resignation the exhausted officer instructed the pilot to turn back again toward Zambia. Further messages had been intercepted that indicated that the supreme commander was injured and still at the scene of the mine blast. It would be a real coup if the team could go in, capture him, and bring him back to Rhodesia for interrogation.

Arriving over Zambian territory for the third time, the team could see two columns of smoke rising out of the hills in the distance. The larger of the two came from the still-burning trucks at the original ambush site and the other from the wreck of the mined Land Rover some two or three miles to the south. Making a beeline for the southern column of smoke, the helicopter pilot suddenly swung his bird around and proceeded to fly back toward the lake. The monitoring unit had intercepted further Zambian Army messages stating that the commander had just died of his injuries.

Apart from being a hell of a tale, there are several learning points to this story. First, it proves just what a well-trained and determined group can do if it is well led, has confidence in its commanders, and takes a bold, innovative approach to overcoming problems (or, as the U.S. Marines say, "adapt, improvise, and overcome"). Second, had the guerrillas employed trackers to check their routes, they would not have been caught with their pants down. Third, the most important lesson of all: *always expect the unexpected!*

Pungi Sticks

Pungi sticks, as all Viet Nam vets know, were responsible for many casualties among U.S. and allied troops untrained in recognition techniques. Sharpened, dung-smeared stakes were hidden in grass and vegetation along the sides of a trail or placed upright in the bottom and sides of a concealed pit to catch the unwary. Strong enough to penetrate the sole of a jungle boot, they produced seriously infected wounds that required immediate evacuation and often took months to heal.

Ambushes

Tragically, tracking teams have walked into ambushes despite using flank trackers as an early warning system. Well-laid and cleverly sited ambushes, even using only one or two men, can be an effective way to terminate a follow-up, especially if casualties are inflicted on the trackers. Therefore trackers should be very wary of the possibility of ambush and trained to recognize and investigate likely ambush sites along the trail. Areas where the team is forced by terrain or vegetation to close up into a single-file formation or where thick brush needs to be entered from open country should be regarded with utmost caution and the appropriate counterambush actions taken. It is far better to find an ambush than to try and fight your way out of one.

Fire

Capt. Allan Savory, founder of the Rhodesian Tracker Combat Unit, had a favorite saying that proved to be right on the money on several occasions. "Never, ever underestimate the offensive use of fire." On two occasions, one deliberate and one accidental, I have been driven off a follow-up due to fire. Once, when initiating contact with a terrorist group, a flank tracker threw a white phosphorus grenade to mark the terrorist position for a circling, rocket-armed spotter plane. A sudden gust of wind—blowing in our direction, unfortunately—created an instant fireball in the tinder-dry vegetation. This drove us back with smoke-inflamed eyes from the searing heat. By the time the fire and smoke had diminished, the lucky group was long gone, and any chance of follow-up was destroyed by the fire. Another time, we had pinned down a wounded but still dangerous terrorist in a large patch of thick bush. It would have been suicidal to go into and search the area for the heavily armed man so, using a pen flare and assessing the correct wind direction, I set fire to the dry grass and leaves hoping to drive him out. In seconds the thicket had become an inferno and a field-expedient funeral pyre for the diehard guerrilla.

Fire, like an antipersonnel mine, is indiscriminate and recognizes neither friend or foe, so if you plan to use it for any purpose, make sure you remain well upwind, have an escape route, and remain in firm control of the situation.

And of course we must not forget our local indigenous indicator!

Other Techniques

A study of unconventional warfare (which is recommended for all would-be trackers) will reveal a multitude of ways to hinder or delay a follow-up. Razor wire, fishhooks, nail boards, and snares can be effectively hidden to catch the unwary. Pit traps, deadfalls, and rockslides have all taken their toll since time immemorial. For an excellent study of this subject, see Ragnar Benson's books *Mantrapping* and *The Most Dangerous Game*, available from Paladin Press.

RECORD KEEPING

When a tracking team or any other tactical unit encounters any new attempts at antitracking, spoor reduction, or countertracking, details of the method used should always be disseminated to other teams and units. Describe the method, the circumstances under which it was used, and if it was successful in stalling or delaying the follow-up. In this way an information file of all known methods is compiled and can be used to brief incoming teams and new trainees.

Tracking is a dynamic and progressive skill, and trackers should always attempt to stay ahead of the game by utilizing a network of information, resources, and experiences. By doing so, they will go a long way toward attaining the elusive goal of tactical awareness.

Chapter 8
COMMAND AND CONTROL OF A TRACKING TEAM

Tactical decisions must be made by the commander on the ground and not by a more senior officer who is not a part of the battle.

Old army axiom

Contemporary U.S. law enforcement and military history is full of cases where highly specialized operations teams have been made to look foolish or, in the worst cases, fail in their missions because they were micromanaged by incompetent superiors. The incidents at Waco, Texas, and Ruby Ridge, Idaho, as well as the aborted mission to rescue captives from the U.S. embassy in Teheran are mute testimony to vitally flawed command and control procedures. The unnecessary deaths and injuries sustained by highly trained Navy SEALs while attempting to immobilize Manuel Noriega's personal aircraft at Paitilla Airport during the invasion of Panama is another example of gross military incompetence displayed by commanders who, despite the availability of better plans formulated by the SEALs themselves, ordered good men into a suicidal situation. (For a detailed account of this tragedy, see Chapter 15 of *At the Hurricane's Eye: U.S. Special Operations Forces from Viet Nam to Desert Storm*, by Greg Walker, published by Ivy Press.)

It is an unfortunate fact of life that many of those who have managed to scramble to the top in the special operations community have done so on the backs of more capable colleagues who spent their time mastering their crafts rather than politicking for influence and power. The point I am making is that trackers, whether police, army, corrections, or special operations forces, *must* be commanded by their own kind of people who know exactly what their capabilities and limitations are. I still bitterly recall the time when my three-man team, wearing only shorts and T-shirts, was ordered to ambush an arms cache on a night that turned out to have the lowest recorded temperature in Rhodesia in 50 years. These thoughtless and inane orders came from a senior police officer who had consumed a large quantity of the local brewery's production and who spent the night in a warm, comfortable bed while we froze our butts off.

Sun Tzu, the Chinese military strategist, wrote: "If you know yourself but not the enemy, for every victory, you will also suffer a defeat. If you know neither the enemy or yourself, you will succumb in every battle but if you know the enemy and yourself, you will prevail every time."

Wisdom that still holds good 2,000 years later.

COMMAND AND CONTROL PRINCIPLES

It is not my intention to go into the politics and mechanics of high-level command and control. Nevertheless, there are some important factors that must be considered in any tactical operation involving law enforcement or military trackers.

In terms of command and control, there are two general types of situations involving the use of tracking teams that require different handling:

1. When the operation is handled by a single agency (e.g., state, county) using its own resources in terms of communications, equipment, and personnel. It is obvious that in a small rural jurisdiction, such operations will probably involve only a small number of officers all known to each other.
2. When it is a multiagency operation conducted by a number of different departments, when providing assistance to a federal agency, or when requiring specific assistance from other agencies. An example of this is a tracking team from one county assisting another county that does not have a team. Such scenarios will bring together officers who are not used to working with each other, possibly having incompatible communication systems. Both will pose cooperation and communication difficulties that, if not handled with maturity and common sense, may lead to operational failure.

Should a multiagency operation be mounted, the following principles of command and control should be followed by all participants if cooperation and cohesion are to result:

- Clear command relationships must be established.
- Responsibilities must be clearly defined.
- Plans and tasks must be stated simply and concisely.
- Tasks and resources must be fairly allocated and distributed.
- Cooperation must be continuous.
- Information must be disseminated up, down, and laterally when and where needed.
- Any changes to operational plans or tasking must be communicated to all participating agencies.

Careful adherence to the above principles will

- ensure unity of effort,
- maintain unity of command,
- determine the best course or courses of action, and
- ensure simplicity in planning and execution.

COMMANDING A TRACKING TEAM

When a tracking team is deployed in the field, a trained tracker must be posted in the operations center who will be responsible for the following tasks:

- Providing advice and recommendations to the commander on team tasks, roles, and capabilities
- Maintaining radio communications between the team and HQ
- Plotting progress of the follow-up on the operations map
- Arranging rotation of teams as necessary
- Arranging resupply as required
- Continually monitoring movement to assess the direction of flight and planning appropriate countermeasures such as observation posts, spoor cutting teams, and ambushes
- Controlling multiple team deployments if leapfrogging tactics are used
- Resisting the C.O.'s desire to use trackers for unsuitable tasks
- Screening the team from the C.O.'s impatience and interference
- Advising the C.O. not to affect the team's

integrity by reducing effective manpower levels (unless the situation absolutely demands it)
- Informing adjoining agencies when trackers on a follow-up cross-jurisdictional boundaries.
- Coordinating and tasking spoor cutting teams
- Determining when to use backtracking teams
- Detailing other available tracking teams with information-gathering tasks

OPERATIONS CENTER REQUIREMENTS

Tracking operations against armed fugitives can only be carried out during daylight hours; under no circumstances should trackers be required to work at night. This means that the tracking advisor in the ops rooms need only be on duty when the team is actively following the tracks, which is generally between first and last light. As long as overnight communications are maintained with the teams in the field, the advisor can stand down and get some rest. In the case of a search and rescue operation, however, it is possible to track at night using flashlights and vehicle headlights.

An operations room can be set up in a corner of the ops center, a small room, a tent, or even the rear compartment of a truck or sport utility vehicle, but it must contain the following items:

- Map board or similar board, about 48 inches x 48 inches
- Clear acetate sheet or something similar to cover the map
- Grease pencils or colored markers
- United States Geological Survey (USGS) map coverage of the operational area, preferably in 1:50,000 scale (1 mile to the inch) or 1:24,000 scale (half-mile to the inch)
- Local information and road maps
- An operations log to record all incoming and outgoing messages and events
- An information file on fugitive's descriptions, footwear, clothing, weapons, etc.
- Radio, antennas, spare batteries, battery charger, etc.
- Cellular phone and directories
- Tracking team information, call signs, medical records, next of kin, etc.
- Information on surrounding jurisdictions and local a law enforcement directory

It is a given that any operation undertaken by more than one agency should be done so in the spirit of cooperation in order to achieve the aims required. Unfortunately, in most cases of interagency activities we see the reverse. Grandstanding, one-man shows, power plays, posturing, jealousy, antagonism, finger pointing, and even outright demagoguery often play a part in combined operations. All of this has no effect other than to cloud judgment and obscure the aim. When men's lives are on the line, the least they can expect from their superiors in the rear is harmony, synergy, and total support of the mission and its goals.

Jack Kennedy once said, "Success has many fathers but failure is a lonely orphan." There are several interpretations of this homily, but the best lesson to be learned from it is that many fathers, working together, guarantees success.

OTHER TASKS FOR TRACKERS

The very nature and training of a tracking team enables it to take on a variety of tasks other than just tracking. Due to its compact size, flexibility, ability to move in wilderness and rural areas with skill and cunning, and its inherent adaptability, commanders have at their disposal an asset that can be used for tasks far beyond the scope of conventionally trained units. These additional missions include the ones discussed below.

Surveillance and Reconnaissance Tasks

Tracking teams attached to drug task forces have the skills to conduct long-term surveillance, particularly in marijuana growing areas. Information on plots, routes used, irrigation systems, and cache areas is vital to planners tasked to interdict or arrest growers.

With their ability to move undetected and remain in remote areas for long periods, tracking teams can provide much valuable information for planning staffs.

It is a remarkable fact that almost half of the marijuana plants destroyed on national forest land in 1993 were from the Daniel Boone National Forest in Kentucky. Park law enforcement officials discovered close to 4,600 plots and destroyed almost a quarter-million growing plants with a street value of almost $250,000,000. In the course of these operations, officers discovered 38 booby traps, including steel-jawed bear traps, pungi sticks, explosive charges, and disguised fishhooks suspended at eye level to catch the unwary.

Narcotics investigators operating in the Ouachita National Forest in Arkansas made 15 felony arrests and destroyed 42,000 plants discovered in 1,940 plots in 1993. As was the case in the Daniel Boone National Forest, a variety of booby traps were deactivated, including one that was wired to a high-powered rifle.

Other national parks that feature high on the "pot list" include the Great Smoky Mountain National Park, the Cumberland Gap Historic National Park, the Shenadoah Park, and a slew of smaller parks along the entire length of the Appalachian Mountains from Pennsylvania to Georgia. Add to this extensive list all the state and federally owned lands in the western half of the country and one can see that the potential ground available for illegal cultivation is vast. It is reliably estimated that marijuana is the largest cash crop produced in the United States. Since this country is the largest producer of agricultural commodities in the world, it is safe to assume therefore that American-grown marijuana must be the largest cash crop in the world.

It is of interest to note that the main areas responsible for growing this enormous quantity of marijuana are also the home to the moonshine industry. Those good ol' mountain boys don't miss a trick, do they?

Search and Rescue

Team tracking skills are not limited to following up fugitives from justice. They can also be used effectively in the location and rescue of missing persons in remote areas. Skilled trackers from the U.S. Border Patrol, for example, have been used with great success in the recovery of lost tourists, children, and hikers.

Livestock Theft

In the western states where cattle ranching is predominant, tracking teams can be used to track stolen and stray cattle. In such cases, trackers mounted on horseback can cover long distances for sustained periods.

Missing Aircraft and Pilots

Every year dozens of light aircraft go missing in rugged terrain, some never to be found. With its ability to operate in wilderness areas, a tracking team is a valuable asset, particularly if the crash site is found and survivors have to be tracked and recovered.

Timber Theft

In the Pacific Northwest, the theft of valuable timber such as cedar and redwood is endemic. (A mature cedar tree is worth several thousand dollars and is a prime target for thieves in the remote, forested areas of the Cascade Mountains.) Trackers are being used to patrol affected areas and react when evidence of illegal activity is discovered. The Department of Natural Resources in the state of Washington has a skilled tracker on staff who has cracked several recent cases of cedar theft.

Gathering Intelligence

The value of good intelligence was amply recognized during both the Viet Nam and Persian Gulf Wars, where small, lightly equipped, mobile teams produced results out of all proportion to their size. Electronic intelligence has its place, but it will never replace a good pair of eyes and ears.

Much useful information can be gained by the correct use of tracking teams. Trained trackers can infiltrate and examine an area, interpret the spoor, and deduce with fair accuracy the events that took place there.

Opportunities for intelligence gathering by tracking teams exist across the country. Several northwestern states have become the stronghold for militant extremists whose activities, motives, and intentions pose a serious threat to local law enforcement and the community at large. Several instances have come to light recently indicating that inner city street gangs have been using national forests to conduct drive-by shooting practice and other illegal activities. A militant Hispanic group known as the Confederate Mexican Army, which openly states that its aim is to restore California to Mexican rule, trains regularly in the remote areas of the Southern Californian desert. A group of Japanese Americans suspected of having links with the Japanese Mafia (*yakuza*) have also been observed conducting paramilitary exercises with firearms and explosives around the Barstow area of Southern California.

Border Patrol

It is an unfortunate fact of life that the U.S. Border Patrol no longer has the time or manpower to track down and apprehend illegal border crossers. Recent investigations in Texas and Arizona reveal that huge quantities of narcotics are pouring into the United States via national and state parks situated along the border with Mexico such as Big Bend Park in Texas. Rangers employed in the Organ Pipe National Monument in Arizona capture about 25 illegal aliens a month and in 1993 recovered almost $4 million worth of drugs, which they estimate is less than 10 percent of the total amount channeled through the park. A well-trained tracking team would almost certainly pay for its training and equipment in less than a week should it be employed there.

Dignitary and Executive Protection

Most heads of state and wealthy executives maintain country estates far removed from the hurly-burly of civilization. Protection for these remote retreats present different problems from those sited in urban environments. A trained tracking team can be used to patrol the perimeter of the estate searching for evidence of the presence of unauthorized individuals or infiltrators. This outer screen of protection can be a critical part of the overall protection plan. Gen. Charles De Gaulle was the target of over 40 assassination attempts while he was president of France, most of them taking place when he was traveling or at one of his country retreats.

In Support of K9 Operations

K9 officers who have attended tactical tracking courses have stated without exception that prior to being taught tracking skills they felt that they were only 50 percent effective, with their dogs doing all the work. As has been pointed out already, dogs have limitations, particularly in dry, dusty conditions. A combination of human and canine abilities will greatly enhance the efficiency of a K9 team, with the human taking over should the dog experience difficulties. (See Chapter 6, Alternative Tracking and Follow-Up Methods.)

ANATOMY OF A FOLLOW-UP

The following scenario will give the reader an idea of how a typical follow-up is conducted from a command and control perspective. Many factors influence a follow-up, and none are alike. There is no set format, duration, or sequence, yet I have attempted to portray two typical situations—one of interest to police and the other for corrections teams—that show events, problems, and procedures as they would occur on a follow-up in the real world.

Scenario

Situation: In a remote, unpopulated area of one of the Appalachian mountain states, a helicopter carrying a detail of National Guard troops deployed on a marijuana eradication task was fired on by four or five men while taking off from an area where the troops had located and destroyed over 500 mature plants. While circling the area the helicopter crew spotted two pickup trucks at a trailhead some 800 yards from the scene of the shooting. The pilot dropped the guardsmen in a clearing close to the trucks, which were assumed to belong to the growers, and ascended to 1,000 feet to communicate the situation to his base station. The base relayed the information to the local sheriff's department, which activated its standby tracking team. The helicopter then flew to the local airport, refueled, picked up the tracking team, and returned to the two pickups, which were being investigated by the guardsmen.

1. The trackers link up with the guardsmen and receive a briefing. The team then examines the area for further information and records the spoor patterns of the alleged growers. A number of sacks and sickles in the back of one of the trucks links the pickup to the growers. The team leader radios the truck tag numbers and descriptions to the circling helicopter pilot, who relays it to the sheriff.

2. Trackers locate the tracks of five persons heading in the direction of the marijuana plots, estimated to be four to six hours old.

3. Tracking controller requests helicopter to relay his SITREP to sheriff. Team call sign is TT; base call sign is TB. SITREP reads as follows:
   Time 10:30
   ALPHA [Location] Grid 464956
   BRAVO [Number] 5
   CHARLIE [Direction] Northwest
   DELTA [Age] 4 to 6 hours
   ECHO [Type] Two heels, plain sole, pointed toe, possibly Western style. One heel, standard military type Vibram cleats. One flat, tennie type, Reebok pattern. One flat, tennie type, edge-to-edge parallel bars, pencil thickness.
   FOXTROT Will uplift by helicopter to area of shooting and investigate. Will advise.
   (*Note that only information of interest to the follow-up is passed back to base*)

4. TB plots above information on the ops map and awaits developments.

5. Helicopter uplifts TT and six guardsmen, leaving two to guard the trucks. Helicopter drops passengers at scene of shooting and lands at a nearby hill to act as radio relay.

6. TT completes 360-degree search of scene. Recovers four freshly fired 12-gauge shotgun shells, six .223 cases, and two .30-30 shells. Relocates tracks of four people leaving the area at a run. One set of tracks unaccounted for.

7. TT controller relays updated SITREP to helo for onward transmission:
   Time 11:05
   ALPHA 456967
   BRAVO 4
   CHARLIE Northwest
   DELTA 8 hours
   ECHO Same but cancel Vibram pattern.
   FOXTROT Evidence of three weapons used. 1 x shotgun 12 ga., 1 x .223 rifle, 1 x .30-30 rifle. Six guardsmen to support follow-up.

8. TT disregards the missing spoor, decides to concentrate on spoor of four, and commences follow-up.

9. TB receives information from state motor vehicle department and passes same to TT through the radio relay. Vehicle owners verified as:
   A. John Starkey, white adult male, age 30, 6 foot 2 inches, black hair, powerful build. Has prior arrests.
   B. Alfred Tucker, white adult male, age 39, 5 foot 7 inches, red hair and full beard, slim build. Has prior arrests

for narcotics and firearms offenses. Considered dangerous.

    C. Known associates: Kenneth Crow and Alfred Garcia. No descriptions at this time.

10. TT relays SITREP to TB:
    Time 12:10
    ALPHA 455982
    BRAVO 4
    CHARLIE North
    DELTA 6 hours
    ECHO Same.
    FOXTROT Located four unfired shotgun shells on trail. Follow-up continues.

11. By this time the TB commander has begun to get an indication of the line of movement of the fugitives and informs TT that he is deploying a spoor cutting team to check a logging road several miles ahead that runs 90 degrees across the anticipated direction of travel.

12. TB relays to TT:
    Time 12:20
    Have deployed TT2 along Old Hope Road four miles to your north to check between locstats 440059 and 470053. Will be in position by 1300.

13. TT plots above information on map and continues follow-up.

14. TT relays SITREP:
    Time 14:40
    ALPHA 451019
    BRAVO 4
    CHARLIE North
    DELTA 4 hours
    ECHO Same
    FOXTROT Fugitives drank water at creek.

15. TB informs TT that TT2 will communicate directly with TT.

16. TT2 to TT:
    Time 15:35
    No spoor found between 440059 and 470053.

17. TT informs TT2 that the tracks are still heading north.

18. TT2 informs TT that no tracks cut along track in the specified search area.

19. Both TT and TB suspect that the fugitives are likely somewhere between TT and the road line examined by TT2. The time/distance gap is closing.

20. TB radios to TT2 to observe the track line where the fugitives are assumed to cross if they continue to move in a northerly direction as they have been doing for the past several hours. TT2 radios its new position to both TB and TT.

21. TT radios updated SITREP to TB, which is monitored by TT2.
    Time 16:05
    ALPHA 456039
    BRAVO 3
    CHARLIE North
    DELTA Cancel 1 Western heel
    ECHO Alphonse Garcia in custody.
    FOXTROT Request helo to uplift suspect.

22. TB transmits permission for helo to uplift Garcia.

23. TT interrogates Garcia about descriptions, physical condition, weapons, and ammunition held by the remaining fugitives.

24. Helo picks up Garcia with two guardsmen as escorts and returns to relay site.

25. TT to TB and TT2:
    Time 17:15
    Three suspects sited about 500 yards ahead.

26. TT2 to TT:
   Have suspects in sight. Moving to intercept.

27. TT2 to TB and TT:
   Time 17:40
   Suspects apprehended at 459049 without incident. Three weapons recovered. Will move and link up with TT.

28. TT and TT2 link up. TT requests TB to send helo to pick up suspects and escort for uplift back to base.

29. TB confirms request and suspects uplifted.

30. Helicopter returns to pick up TT and TT2 for return to TB.

Comments

Although this is a straightforward scenario, it shows how timely information passed back to the operations center enables the commander to stay abreast of the tactical situation and deploy his resources accordingly. The use of a relay station, in this case the helicopter, may be a necessity in mountainous terrain to facilitate the flow of radio traffic between control and tracking teams. Also note how little time is spent on the radio and the lack of interference by command and control during the follow-up. Messages sent during the follow-up are always according to the ABC method with *accuracy, brevity,* and *clarity*—no more, no less.

The commander at base must monitor the follow-up, be continually aware of any requirements that could arise, and plan in advance to cope with every possible contingency. What must he do, for example, in the event of a casualty in the tracking team or follow-up support group? What does he do about a resupply of radio batteries? How is he going to get water to the follow-up team? All of these requirements must be taken into consideration, planned for, and rehearsed. Suitable standard operating procedures (SOPs) must be prepared so that nothing is overlooked that could negatively affect the follow-up while in progress.

COMMAND AND CONTROL EXERCISE

The following command and control exercise, to be conducted in a classroom, is designed to give insight into the procedures to be used on a follow-up and to provide tracking team members with some practical problems to overcome.

All that is required are two radios, some message pads, and several maps of the exercise area, which in this case is the United States Geological Survey (USGS) map, KNIGHTS FERRY QUADRANGLE 1859 I SW. Scale 1:24000 (approximately 2.5 inches to the mile).

A person representing the tracking team sits outside the classroom and transmits the radio messages in sequence to an operator, representing the control room, who remains inside. The trainees record each message as it is sent and mark their maps according to the information relayed. In this way they will record on their maps the progress of the follow-up and become aware of some of the problems and answers that will surely occur in a real situation. The team leader or instructor can stop the exercise at any time and question the trainees on what they would do if they were in control of the operation or to suggest alternative courses of action.

This exercise, representing a two-day situation in real time, should take no longer than an hour to complete in the classroom.

Situation: A national forest in California's Sierra Nevada foothills. 0810. A work party of eight inmates from a nearby state corrections facility has overpowered and tied up its guards and absconded in a Forest Service 4x4 Jeep Cherokee. One of the guards has managed to loosen his bonds and release his co-workers. Another younger and fitter officer has run more than six miles to a fire tower and contacted the regional forestry officer on the tower radio. In accordance with SOPs, the corrections authorities have informed the state highway patrol and local and county sheriffs' and police departments, alerted their special response team (SRT), and opened up their operations center.

Command and Control of a Tracking Team

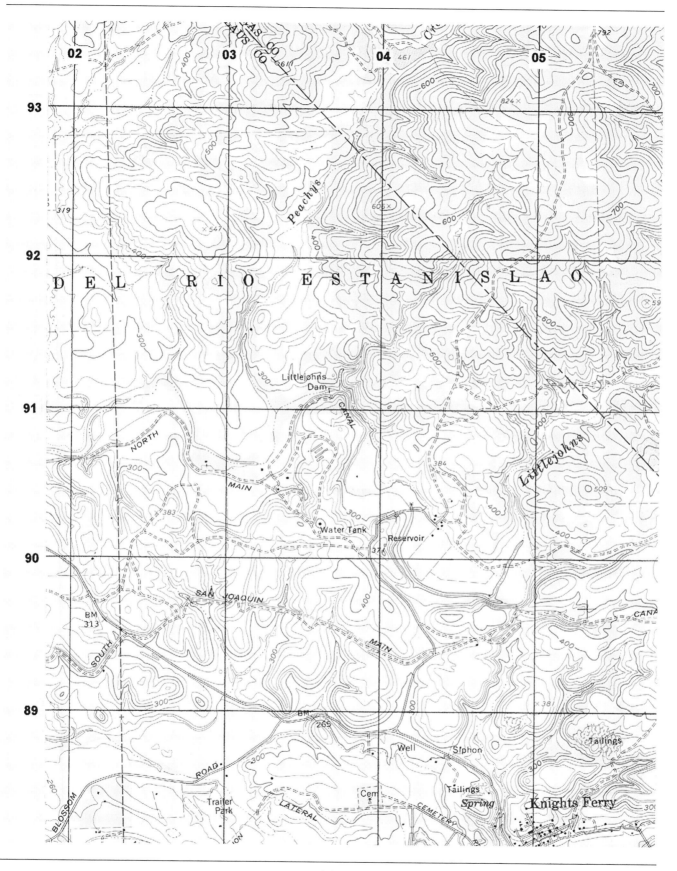

Tactical Tracking Operations

1. 0945. Operations center issues names and descriptions of the escapees. Three white, four Hispanic, and one black, all wearing green corrections overalls, baseball caps, and identical footwear with standard Redwing sole patterns. They are known to have in their possession two Remington Model 870 12-gauge shotguns with five shells per gun and two .38 Special Smith & Wesson revolvers with six rounds each, plus two corrections department multi-channel radios with battery capacities of 12 hours each.

2. 1100. Gravedigger working at a cemetery reports an apparently abandoned, out-of-gas Forest Service Jeep at 039883.

3. 1105. Operations center dispatches SRT with trackers to cemetery area.

4. 1145. SRT (call sign SF1) arrives at cemetery and examines scene.

5. 1200. Message from SF1 to ops center:
   ALPHA 039883
   BRAVO 8
   CHARLIE West along track
   DELTA 4 hours
   ECHO 4 Redwing
   FOXTROT On tracks and moving

6. 1232. From SF1 to ops center:
   ALPHA 039885
   BRAVO No change
   CHARLIE No change
   DELTA No change
   ECHO No change
   FOXTROT Tracks leading to trailer park

7. 1248. From SF1 to ops center:
   ALPHA 029886
   BRAVO No change
   CHARLIE Northwest
   DELTA No change
   ECHO No change
   FOXTROT Track skirted trailer park, crossed Orange Blossom Road

8. 1310. From SF1 to ops center:
   ALPHA 022891
   BRAVO No change
   CHARLIE Northeast
   DELTA No change
   ECHO No change

9. 1315: State police call ops center with a report that a rancher heard radio report of escape and claims he saw about six men in green clothing crossing the road at 021899. Control passes this information to SF1, which abandons tracks and moves in own vehicles to scene of sighting. Rendezvous with rancher, who indicates tracks.

10. 1340. From SF1 to Ops Center:
    ALPHA 021899
    BRAVO Confirmed tracks of eight crossing road
    CHARLIE Northeast
    DELTA 4 hours
    ECHO No change

11. 1505. From SF1 to ops center:
    ALPHA 028905
    BRAVO No change
    CHARLIE East
    DELTA 4 to 6 hours
    ECHO No change
    FOXTROT Tracking difficult due to cattle movement

12. Note: At this time, no specific direction of travel has become evident, so the ops center cannot utilize other available assets at this time.

13. 1610. From SF1 to ops center:
    ALPHA 033904
    BRAVO No change
    CHARLIE Northeast
    DELTA 6 hours
    ECHO No change
    FOXTROT Line cabin at 033905 broken into and ransacked. Request you contact owner and inform me of anything taken by fugitives that could affect follow-up.

14. 1735. From SF1 to ops center:
    ALPHA 036907
    BRAVO No change
    CHARLIE Northeast
    DELTA 8 hours
    ECHO No change
    FOXTROT Lost tracks due to heavy cattle movement and poor light. Will remain overnight and continue first light tomorrow.

15. At this point, the operations controller makes an appreciation of the developing situation and decides to send a reserve tracking team at first light to cut for tracks from a point at 051930 south along the road to junction at 043909. Also accepts offer of use of state highway patrol fixed-wing aircraft and helicopter on standby.

16. 2025. From SF1 to ops center:
    Single shot heard approximately 2 miles east of my position.

DAY TWO

17. 0545. From SF1 to ops center:
    ALPHA 039907
    BRAVO 8
    CHARLIE South
    DELTA 12 hours
    ECHO No change

18. 0610. From ops center to SF1:
    Have sent call sign SF2 to your north to cut for tracks from 051930 to 043909.

19. 0710. From SF1 to ops center:
    ALPHA 044903
    BRAVO 5
    CHARLIE North
    DELTA 12 hours
    ECHO No change
    FOXTROT Found remains of a calf shot and butchered at abandoned building. Tracks of five going north and three going southeast. Will continue with northerly tracks.

20. Ops center informs SF1 that SF2 will move south and follow up breakaway tracks of three going southwest.

21. 0820. SF1 to ops center:
    ALPHA 044903
    BRAVO 3
    CHARLIE South
    DELTA 4 hours
    ECHO No change

22. 0835. SF2 to ops center:
    ALPHA 041916
    BRAVO 5
    CHARLIE North
    DELTA 6 hours
    ECHO No change

23. 0900. From ops center to SF1:
    Have uplifted two observation teams by helicopter. Call sign OP5 at 029992 and OP6 at 041923.

24. 0915. From SF2 to ops center and SF1:
    ALPHA 0489898
    BRAVO 3
    CHARLIE South
    DELTA 4 to 6 hours
    ECHO No change
    FOXTROT Tracks heading straight to Knights Ferry. No attempt to hide or evade. Fugitives using cattle paths and trails.

25. 0940. From SF1 to ops center:
    ALPHA 035927
    BRAVO 5
    CHARLIE Northeast
    DELTA 4 hours
    ECHO No change
    FOXTROT Tracks suddenly swung northeast, showing evidence of running and hiding.

26. 0943. From ops center to SF1:
    OP5 was dropped off at 0815 by helicopter, possibly causing fugitives to hide. Will uplift OP6 to cut for tracks on road between 022933 and 041934.

27. 1025. From SF1 to ops center:
    ALPHA 039932
    BRAVO 5
    CHARLIE Northwest
    DELTA 4 hours
    ECHO No change

28. 1040. From ops center to SF1:
    OP6 dropped at 022933. Have informed sheriff of Calaveras County you are now in his jurisdiction.

29. 1120. From SF1 to ops center:
    ALPHA 031937
    BRAVO 5
    CHARLIE Southwest
    DELTA 4 hours
    ECHO No change
    FOXTROT Tracks on south side of road

30. 1150. From OP6 to ops center:
    Have completed sweep of road between 022933 and 031937. No tracks found. Have linked up with SF1.

31. Note: Dropping OP6 by helicopter to search the road ahead of the follow-up team has caused the fugitives to swing south away from their intended direction.

32. 1325. From SF1 to ops center:
    ALPHA 022932
    BRAVO 5
    CHARLIE Southeast
    DELTA No change
    ECHO No change

33. 1240. From ops center to all call signs:
    Three fugitives surrendered to deputies in Knights Ferry at midday. Interrogation reveals that balance of group remains armed and ready to fight to avoid capture. Also reveals that one fugitive has a relative living in the trailer park at 028015. [Note that this is the trailer park referenced in step 7 of this exercise.]

34. 1345. From SF1 to ops center:
    ALPHA 034927
    BRAVO 5
    CHARLIE South
    DELTA No change
    ECHO No change
    FOXTROT Tracks show evidence of running

35. 1500. SF1 to ops center:
    ALPHA 031914
    BRAVO 5
    CHARLIE South
    DELTA No change
    ECHO No change

36. 1620. SF1 to ops center:
    ALPHA 032904
    BRAVO 5
    CHARLIE South
    DELTA No change
    ECHO No change
    FOXTROT Fugitives walked in stream in an attempt to antitrack

37. 1710. SF1 to ops center:
    ALPHA 028898
    BRAVO 5
    CHARLIE South
    DELTA 2 hours
    ECHO No change
    FOXTROT Shotgun found in canal

38. 1725. Ops center to SF1:
    Assessed that fugitives will attempt to make contact with relative in trailer park to your south. Have placed call sign AB1 in area. Call when within one mile of area.

39. 1800. SF1 to ops center:
    Light conditions poor so will rest overnight at 028891 north of trailer park. Tracks very fresh, about one hour old.

40. 2100. SF1 to ops center:.
    Shots heard from south of my position.

Presumed to originate in area of trailer park. Please advise.

41. 2105. From ops center to all call signs: Remaining five fugitives arrested by AB1 in area of trailer park after brief exchange of fire. No casualties. SF1 to move to trailer park ASAP for pick-up tonight.

42. Exercise ends.

Note that although the operations controller was initially unable to ascertain a positive direction of flight in order to deploy his other resources, he used both spoor cutting patrols and observation posts effectively. The use of these assets prevented the fugitives from breaking out of the area as well as caused them to turn away from their intended destination, thereby giving the initiative to the pursuers. The steady movement of the fugitives in a southerly direction on day 2 led the controller to correctly deduce that the fugitives were heading for the trailer park where it was known one of them had relatives. He then placed an ambush on the trailer, which resulted in the capture of the group without casualties on either side.

Summary of the Main Points
1. Use teams according to their training and abilities.
2. Do not harass tracking teams with unnecessary radio traffic.
3. Use spoor cutting teams when the general direction of flight becomes apparent.
4. The wise use of both fixed-wing and rotary-wing aircraft can put considerable pressure on fugitives and cause them to panic and make unsound decisions that will lead to their capture.
5. Use leapfrogging tactics when necessary.
6. Use all sources of information to build up a picture of the fugitives, their capabilities, and possible intentions.

Chapter 9

HIGH-TECH HELP FOR TRACKING TEAMS

Better to have it and not need it than to need it and not have it.

Old soldier's saying

As stated earlier in this book, tracking is a human skill in which nothing can replace human eyes, ears, nose, touch, and intuition. There are, however, several pieces of high-tech equipment on the market that can assist the tracker significantly by saving him time and effort and enhancing his human faculties.

Anything that enables the tracker to save time or escalate the speed of the follow-up is valuable and could make the difference between success and failure. The important point to remember is that extra equipment must be portable, reliable, and user friendly and not materially increase the weight of a tracker's basic equipment load. There is always a point where the operational value of any piece of gear is countered by its weight or inconvenience. The trick is to experiment to see exactly where that point is.

During the early days of my military career we had a unit transport officer affectionately nicknamed "Desert Toad" who epitomized the point I am trying to make here. Dear old DT would never use his load-bearing suspenders, preferring to fix all of his gear directly onto his pistol belt. Being a fairly rotund sort of guy with lots of belt space, he could cram on his pistol, compass, two canteens, machete, magazine pouches, first aid kit, knife, rations, pen flare, and other sundry items until he looked like a Christmas tree in a surplus store. Everybody would watch him until, strained beyond the limit, his belt would break, dumping all his gear onto the ground.

DT never learned—he did this every time he ventured out on ops, and it provided us with a constant source of amusement. The point is, find out what works, evaluate its usefulness to the mission at hand, and make an informed, intelligent decision as to whether its usefulness is worth the extra energy you will expend carrying the darned thing.

That said, items that demand serious consideration are these:

- Global positioning system
- Infrared sensors
- Multiple-use binoculars
- Multiple-function watch
- Listening devices
- Lightweight communication systems

GLOBAL POSITIONING SYSTEM

A global positioning system, or GPS, is an essential piece of equipment these days, particularly if the follow-up takes place in remote, rugged terrain where normal navigation skills are tested to the extreme. There are several models currently available that retail between $200 and $700, which is not a high price to pay for instant location information.

The GPS made its debut during the Gulf War when soldiers operating in featureless desert terrain needed artificial aids by which to navigate. Although the sun compass, which had been the only method available for years, was satisfactory if a bit slow, it needed much practice and experience to be used accurately. Due to the fast-moving nature of mobile warfare, the slow sun compass had become obsolete, and immediate, real-time navigation plots were essential to direct both men and gunfire to achieve mission requirements. Using in-place satellites for triangulation and coded radio signals, the GPS was designed to provide immediate navigational data to units anywhere in the world. Fortunately this technology is now available to the general public.

Of interest to the tracker is the hand-held, lightweight model utilizing 24 orbiting satellites that transmit their position continually. The GPS picks up a number of these signals and quickly computes its own location to within less than 60 feet. The system has the ability to plot a position anywhere in the world within two minutes of initial activation, and updates take as little as a few seconds to compute. Signals are normally computed and displayed in latitude and longitude, but several models have the ability to convert to the standard UTM grid used on USGS maps, which a tracker would more than likely use on operations. From the tracker's point of view, it enables him to follow a set of tracks without having the bother of continually reading his map to ascertain his position prior to sending off his SITREP. This is of enormous value in terms of saving time, but it must be remembered that the GPS *does not* substitute for good map reading skills.

Some of the features of the GPS are as follows:

- Gives position in latitude and longitude.
- Can convert to grid references on the UTM system.
- Guides both day and night, in fog or blizzard.
- Can calculate distances and azimuths from point to point.
- Sets up to 100 waypoints to assist cross-country movement and can guide you back to your start point.
- Tracks in miles, kilometers, or nautical miles.
- Calculates elevation.
- Operates between 14 and 140° F.
- Is waterproof and ruggedly constructed.
- Weighs less than a pound.
- Uses universal AA batteries.
- Has a battery life of up to 10 hours of use.

Used specifically in the tracking or search and rescue role, a GPS can do the following:

- Quickly establish the follow-up commencement point, which is essential for the command and control element to begin their map plotting.
- Regularly update command and control of the team's location during the follow-up.
- Establish accurately any specific point along the follow-up route.
- Plot water points, caves, hiding places, or other places of possible intelligence value for the future.
- Enable the team to advance quickly and accurately if the tracks are found by a spoor cutting team.
- Relocate specific points along the route at a later date if required.
- Guide the team to a pickup or resupply point and then back to the trail.
- Record positions of any booby traps or mines for later disposal by EOD teams.
- Guide in replacements or support teams.

As you can see, the GPS is of tremendous

value to the tracker. You can never be lost, you can retrace your steps, you can rendezvous with others, you can relay your exact position to a searching aircraft, and you can navigate to any known position with up to 15 legs. However, all this is true *if* you have an instrument that is in working order, have not dropped or lost it, have maps to work from, carry sufficient batteries, and have been correctly trained in its use and understand all its functions. It must be repeated that instruments of this nature are advantageous, but they do not replace good map reading and navigation skills. It would be sheer folly for a team to deploy on a follow-up relying solely on a GPS.

Two models of interest to a tracking team are these:

- *Trimble Scout GPS* (MGRS—Military Grid Reference System). Made by Trimble Navigation, it costs about $700 and is available from the U.S. Cavalry Store.
- *Magellan Recon Trail Blazer* (MGRS). Manufactured by Magellan Systems Corporation, it is available for around $459 from the U.S. Cavalry Store.

The GPS offers a variety of advantages to a tracking team engaged on a follow-up. But be warned: it is only as good as the life of the battery! Take plenty of spares.

INFRARED SENSORS

Another valuable addition to a tracker's inventory is the infrared (IR) sensor. Resembling a medium-sized flashlight, it consists of supersensitive, computer-assisted software and state-of-the-art military IR technology. Weighing less than half a pound, this hand-held device will detect a heat source under any conditions, even while on the move. After extensive testing by an Idaho sheriff's department, it was found to be effective at up to 3,000 feet in open country and 450 feet in wooded terrain. The device works by sensing the temperature change between an animal or human target and the ambient temperature. It can pick up warm engine blocks, hidden fires even when extinguished, and human bodies up to 24 hours after death. Line of sight is not required, since it is designed to work over a 90-degree arc, and it will penetrate brush, trees, smoke, mist, and rain.

An IR device is ideal for finding concealed fugitives under the following conditions:

- When the tracks are fresh and contact with the quarry is imminent
- When your gut instinct tells you that the quarry is close by
- In searching areas of thick brush with reduced risk to the tracking team
- In searching for wounded fugitives when following blood spoor
- In scanning for latent heat sources as part of lost spoor procedures
- In seeking hidden fires at night from a high vantage point (including aircraft)
- In scanning likely ambush sites during the follow-up

There are currently three models on the market, all manufactured by Game Finder, Inc. of Huntsville, Alabama, and available directly the manufacturer or from the U.S. Cavalry Store. Prices range from $200 to $500, depending on model chosen.

- *Game Finder GFI-IR1*. Weighing 6.1 ounces, it is the least expensive model in the range. It is powered by a nine-volt battery that has a life of 12 hours. It is reported to be able to detect a deer up to 100 yards on a 65°F day and a hidden man at 150 yards. Price is $199 from the U.S. Cavalry Store.
- *Game Finder GFI-IR2A*. At 6.3 ounces, this detector has both manual and automatic modes of operation and continually adjusts itself to prevailing conditions such as ambient air temperature, wind, humidity, type and size of heat source, as well as IR noise and clutter. It has the additional advantage of superior motion detection and automatically selects and indicates the most significant heat sources. Price is $300 from the U.S. Cavalry Store.
- *Life Finder*. This is the top-of-the-line model

specifically designed to locate hidden humans and will even locate missing body parts. Weighing in at only 6.3 ounces, it comes with a pouch and instructional video. It works, as do the others, for up to 12 hours on a nine-volt battery. This tactically advanced instrument is warranted and insured for law-enforcement personnel and is a critical must-have in life-threatening follow-up situations. But remember to keep a resupply of batteries on hand—its 12-hour life is not a long time when engaged on a follow-up. Keep it turned off when not in use. Price is $499 from the U.S. Cavalry Store.

Heat-sensing devices are already playing a major role in locating hidden criminals, particularly at night. As a tracker's tool they can be the deciding factor in making successful contact with the fugitives and may save officers' lives. It must be emphasized that to become proficient with an IR sensor, the operator must put in a lot of practice time.

MULTIPLE-USE BINOCULARS

Another essential piece of equipment for a tracker is a good pair of binoculars. There are dozens of models, sizes, styles, and magnifications to choose from, but modern technology has again come to our assistance in manufacturing binoculars with more than one use. Tasco offers an excellent 7 x 50 model with a built-in compass and range finder. Built for use under any weather conditions, it has a waterproof and fogproof rubber armored body. The optically clear and bright 7-power lens brings the viewed object seven times closer. The 50mm objective lens has excellent light-gathering qualities that improves viewing under low-light conditions. Containing its own illumination source for both the compass and range-finding reticle, these binoculars can be used both day and night. Antireflection shields can be fitted to cut out the possibility of observers picking up sunlight reflecting off the lens. At a little over two pounds in weight, they retail for around $200.

If two-lb. binoculars seem unnecessarily heavy, also available is a 12-oz. monocular called the Data Scope. This precise, compact instrument enables the user to find an azimuth, measure distances, and keep the time all at once. By looking through the lens and pushing a button the operator has at his fingertips the ability to select either compass, range finder, or chronometer modes. The Data Scope's digital reticle superimposes the bearing, range, and time data right on the field of vision. It is shock resistant, waterproof, and a mere four inches long. The Data Scope is powered by three lithium batteries, so don't forget to take along a supply of replacements.

The ability to see clearly at long distances or through thick brush is critical to the tracker, so the choice of a good vision-enhancing instrument is very important. The field is wide open, with hundreds of models to choose from, but if you can get three functions for the price (and weight) of one, it makes good sense to get hold of a multifunction set. Even if the batteries do run out, the binocular feature will still be functional.

MULTIPLE-USE CHRONOMETERS

I have a $3,000 Rolex watch. Apart from being a nice piece of jewelry, all it does is tell the time and no more. Currently on the market is a $130 digital watch from Casio that not only tells the time but provides compass readings, an abnormal field indicator, five memories, alarm, stopwatch, countdown timer, calendar, microlight, and more. You think this is good? Well Casio also has an even more remarkable high-tech version. For $200 you get a chronometer that not only tells the time, it gives an azimuth or compass bearing, your altitude, the temperature, the barometric pressure, and two graphs: one to alert you to changing weather conditions and the other indicating ascent and descent data. It has five different alarms, a stopwatch function, calendar, and a light to make it visible at night. This amazing piece of electronic wizardry stores up to 50 sets of data. And all from a watch!

I suspect you will need a master's degree in electronics to be able to use all the functions, but given the choice I would take the Casio over the Rolex any day. With its multitude of functions, particularly those related to navigation and map reading, it is a must for any tracker, indeed any soldier or police officer, and will save a lot of time as well as make other, more traditional items of equipment redundant. (Don't forget to take a spare battery!)

SOUND ENHANCEMENT DEVICES

Human hearing can be hindered by many factors, including wind, background noise, bird and animal calls, distant traffic and aircraft, as well as physiological problems like hearing loss or damaged eardrums. From a tracking point of view, all of the above factors could affect a tracker's ability to detect noises that could signify a hidden danger. Not to worry, though. There are several devices on the market that enhance distant sounds as if they originated only a few feet away. Under certain circumstances the ability to locate and interpret distant sounds could be of major benefit to the tracker, particularly if the state of the tracks indicates that the quarry is nearby.

One such device is the Bionic Ear, which magnifies sound beyond the human range. A whisper can be heard at 50 yards, as can people walking up to 1,000 yards away. When fitted with an optional dish-shaped directional booster, efficiency increases by over 30 times and cuts out background clutter. The basic units retail for around $100; the booster dish goes for $30.

Another advanced hearing system has just been introduced by an innovative company in Pennsylvania called Walkers Game Ear. Known as the Tact'l Ear, this amazing hearing enhancement/protection device fits into the ear in the same way as a hearing aid. With a battery life of 140 to 180 hours, it weighs less than an ounce. Advanced technology allows for omnidirectional hearing with emphasis on high-frequency sounds such as muted or low-level conversations, loading or cocking of a weapon, the clicking of a gun's safety catch, and footsteps. Not only does this little device enhance hearing six times, it also provides maximum hearing protection. A special safety circuit shuts off the Tact'l Ear when a firearm is discharged, thereby giving added protection to the sensitive hearing organs.

The first time I put this wonder in my ear I was amazed to hear sounds that my tired old ears had not heard for years. Even jingling change in my pocket sounded like a bus passing by! The Tact'l Ear is a highly recommended piece of equipment that will give its wearer a decided advantage in any tactical situation and enhance officer safety and survival. Bob Walker of Walkers Game Ear offers a special price to law enforcement officers of $219.95 (usually $299.95). This price includes a well-made black Cordura pouch by Uncle Mikes of Oregon.

From an operational point of view, a combination of a listening device and an IR sensor used in thick, dense brush is tactically and technologically sound and will probably assist you in pinpointing fugitives fairly accurately, especially if they are unaware of your presence and continue to act in a normal fashion.

ULTRALIGHT COMMUNICATIONS EQUIPMENT

Recent developments in microcircuit technology have decreased the size of personal communications apparatus to the point where they are practically weightless. A tracker on the spoor will be more effective if he has the ability to remain in constant communication with his controller and flank trackers so that when developments dictate, information can be passed on immediately.

Currently under trial are new radio headsets that are surgically implanted into the skull behind the ear. It remains to be seen whether they function under the rigors of operational use, but it does give you an idea which way communications technology for police and military special operations community is heading.

It must be emphasized that any headset worn by a tracker must not impair his natural hearing ability in any way, so care must be taken to provide equipment that is functional without being distracting.

Television Equipment Associates (TEA) manufacture several headsets suitable for tracking operations:

- *LASH I*. This headset consists of a single earpiece with a soft ear tip held comfortably in place by a plastic ear curl. A principal feature of this model is that it allows the user peripheral hearing, a plus for tracking work. It comes with a throat mike with breakaway strap. The microphone is water resistant, and voice transmission remains the same irrespective of the user's actual voice volume (usually several octaves higher when the bullets begin to fly!) so that a whisper will be transmitted as normal speech. For special orders, TEA can provide a thumb switch located on the end of a thin cable that can be run down the operator's arm.
- *LASH II*. This is an improvement on the LASH I and consists of a throat strap with a microphone and receiver/speaker. The speaker transmits sound to either a molded ear plug or an earpiece secured in place by a clip. The result is an inconspicuous and efficient unit. Both of the above units are designed to be used with any standard police radio.

Maxon manufactures a neat little system with a voice-activated circuit. The adjustable headset features an earpiece, boom mike, and built-in whip antenna. Retails for about $40.

Sonic Communications, an innovative British company with an office in the United States, produces a compact radio ideal for tracking use known as the Sonic Ear Microphone. Taking advantage of microcircuitry, this unique little communications device combines the transmitter with the receiver. The earpiece is placed into the ear in the normal way for receiving messages, but when you speak it becomes a transmitter. This neat, compact, and functional radio fulfills just about every tactical need and is ideal for tracking team communications.

(Note: Voice-activated units are not recommended for trackers engaged in pursuit of armed and dangerous fugitives. There is a possibility that unwanted sounds may be transmitted, particularly if the tracker is panting due to exertion. This could give away his position to a hidden fugitive and give the latter the tactical edge.)

CELLULAR PHONES

Another example of modern technology that has become universal in the past few years is the cellular telephone. What a boon to any operational field officer to extract his pocket phone, flip it open, and, like ET, call home! The advantages to the tracker on a follow-up are enormous. He can quickly and easily contact and be contacted by just about anyone in the command and control channel. He can pass information with total clarity and not have to worry about static and atmospherics, which were the bane of regular radio communications. If, however, you are in the act of sneaking up on your quarry, be sure to turn your phone off so that if some unthinking rear echelon individual chooses to call you at that moment it will not give you away!

In many ways, the cell phone has made HF and VHF communications redundant, and with its touch-tone capability, good connections are virtually guaranteed. Don't leave home on a follow-up without one, but don't forget your directory and a resupply of batteries too.

SUMMARY

It must be remembered that the aids outlined in this chapter are just that—aids. They do not replace real human skills and abilities. There is no rule that says you cannot use gadgets, but you have to consider the trade-off between the benefits they will provide and the additional weight you will have to carry. Also consider that in earlier times, and not so far back either, trackers carried out very successful follow-ups without the benefit of any of these devices. So these aids are not necessarily critical to success, although there is no doubt that they can make a tracker's job that much easier and even mean the difference between success and failure. It's your call!

I opened this chapter with a wise old saying that should never be forgotten by any law enforcement officer or soldier: "Better to have it and not need it than to need it and not have it." I would like to add to that pearl of wisdom: "If you take it, you have to carry it!"

Chapter 10
DEVELOPMENT OF A TRACKING TEAM

They're only puttin' in a nickle, but they wanna dollar song.

Song title

You would not be reading this book unless you are interested in the subject of tracking. Chances are that you are not the run-of-the-mill cop or corrections officer. It's more than likely that you are the sort of person who volunteers for special duty, are a member of a SWAT or SRT, or have an above average interest in unconventional law enforcement or military operations. You obviously feel that tactical tracking may be a subject of value to your jurisdiction, department, or facility. This chapter covers some of the factors, both positive and negative, that may have to be taken into consideration by you, your superiors, or your colleagues when discussing the need for establishing a tracking team or even getting the funding required to have some of your people trained as trackers.

Apart from the fact that every K9 officer will benefit from tracking training, not every police agency will require the services of a dedicated tracking team, even if they do maintain an SRT. However, it is highly recommended that corrections facilities on both state and federal levels institute such a resource. Trackers, generally speaking, operate in rural, remote, or wilderness areas. Having said this, most jurisdictions, even major cities, have parks and open spaces that draw the criminal element like flies to a trash dump.

There are, however, many times when a tracking team can be of great benefit to any jurisdiction. It would be an important element in searching for evidence at a crime scene or conducting a search-and-rescue operation. When the U.S. Border Patrol still maintained trackers on its books, particularly in the Southwest, they were often employed in the search for lost children or tourists with great success. (Jack Kearney, author of the book *Tracking: A Blueprint for Learning How,* personally found more than 60 missing persons.)

It is often said these days that tracking is a vanishing art. This need not be so, especially if a successful search-and-rescue operation involving the use of trackers captures the attention of the sensation-hungry media. Trained trackers are already recovering escaped convicts in the state of Washington, and a federal marshal once told me that in his opinion, the Ruby Ridge disaster might never

Trackers searching for spoor shortly after an air strike on a guerrilla base camp. In this instance, these four trackers, by following blood spoor, were able to account for a number of wounded guerrillas who had managed to hide in the surrounding dense brush.

have happened had the officers involved received instruction in tracking techniques.

From a military standpoint, a man trained in tracking operations, even if he is not a master of the art, will make a much better soldier as a result of his exposure to new ideas and techniques. As the commanding officer of the Rhodesian Tracking School, I was often informed by other officers that the skills and knowledge gained by soldiers on a tracking course were of inestimable value during small unit counterinsurgency operations.

The question to ask is, "Does my jurisdiction (or institution) need a tracking team? Do we have the manpower, time, and money for training and operations?"

Perhaps this question can be answered by the following comments made by an investigator from a major penitentiary: "I have investigated escapes from the institution on several occasions, and now that I have received tracking training I believe that my ability to gather timely and pertinent information about the escape has been greatly enhanced. I am now able to better determine avenues of escape, direction of travel, approximate time of escape, and rate of travel, all of which will benefit the department's ability to apprehend escapees and identify probable avenues of future escape attempts. In time, with a better understanding of tracking operations, we will be able to better safeguard the public with a swift recovery of any escaped felons that could threaten their safety." I think that sums up precisely why every corrections facility should have a tracking capability, even if it is only two or three men.

FACTORS TO BE CONSIDERED

If your jurisdiction has several of the following conditions, a tracking team would be a valuable addition to its operational resources:

- Vast, uninhabited, or sparsely populated areas
- Terrain suitable for the employment of tracking teams
- Availability of trained manpower with local knowledge
- Areas known to be supporting or suitable for marijuana cultivation
- A seasonal influx of tourists or special interest groups
- Ideal hiding places for drug labs
- Wilderness areas attracting hikers and campers
- Areas already used for illegal training by hate groups or criminal gangs
- A county, state, or federal corrections institution
- Cult or survivalist compounds nearby.

- A history of illegal activities (e.g., narcotics, moonshine)
- Timber thefts
- Acts of environmental terrorism
- Spotted owls or other endangered species
- A nuclear power station
- Sensitive or controversial environmental sites
- A history of toxic waste violations
- Sensitive military facilities

In order to establish an effective tracking team, several things must happen:

- Team members must want to be team members and be eager to learn, willing to train, and have a desire to excel.
- Team leaders must have command experience and a strong knowledge of tactics.
- The administration must support the program wholeheartedly.
- Funds must be made available for training and equipment.
- Training time must be made available.
- Facilities must be made available for training.
- Training must be mandatory, professional, and documented.

The most elite special operations units in the world are only successful because all of the above conditions have been fulfilled. There is an old military maximum that says, "You cannot operate successfully in the field unless your rear base is secure." Although it is not expected that administrations will have on hand all of the necessary funding, this is only a small part of the entire picture. Reasonable funding and allocations of training time, combined with a strong desire on the part of the team, will ensure the agency of a highly trained and proficient asset.

Flexible, mobile, and versatile, four-man tracking teams can be used on a variety of tasks other than following and locating fugitives. This tracker is checking a well-used power line service road for evidence of guerrilla-emplaced antivehicle mines in Namibia.

Trained police trackers deployed on a drug eradication program discuss suspicious tracks found in the pine forests of the Pacific Northwest. The temperate climate and remote forests provide excellent conditions to cultivate marijuana. Tracking is a cost-effective method to locate and destroy growing plants and provide good intelligence on routes used by growers.

Police trackers follow the spoor of marijuana growers from plot to plot through dense brush in a forested area in the Pacific Northwest.

EQUIPMENT

Suitable equipment and weapons will play a vital role in the team's ability to carry out a successful follow-up. Like it or not, cops are "gadget oriented," so care must be exercised not to waste limited finances on Star Wars technology that will seldom if ever get used, either because it is too technical to figure out or because the need for it seldom arises. However, though it is true that a person does not become a carpenter when you give him a hammer, it is also true that a proficient carpenter can work more quickly and efficiently if he has one. Fancy equipment in no way substitutes for good training, but properly selected equipment in the hands of a well-trained team can increase their odds of success.

The basic support equipment needed by each team is as follows:

1. Radio with earpiece and remote (or throat) microphone
2. Compact first aid kit
3. Ground/air code panels
4. Binoculars
5. Satellite global positioning system (with UTM conversion capability)
6. IR heat sensor
7. Signal flares or pencil flares
8. Strobe light
9. Maps of the operational area.

Each team member must have the following:

1. Primary and backup weapon
2. Multipurpose tool (e.g., SOG, Leatherman)
3. Load-bearing equipment
4. Flexicuffs
5. Uniforms, woodland camouflage or similar, boonie-style hat
6. Suitable comfortable footwear, preferably boots
7. Dark socks
8. Belt, military webbing type
9. Camouflage face cover/sniper veil
10. Brown or camouflage t-shirt
11. Backpack
12. Holster (if handgun is carried)
13. Magazine pouches
14. Four canteens and pouches
15. Rain poncho
16. Fixed-blade or lock-back knife
17. Waterproof notebook and pencil
18. Watch, preferably luminous
19. Compass and protractor
20. Small flashlight (e.g., Mini-Maglite)
21. Roll of paracord
22. Green tape (for silencing equipment)
23. Food supply
24. Water-purifying tablets
25. Ammunition and pyrotechnics

As can be seen, there is very little over and above the gear already possessed by most members of an SRT. The acquisition of a GPS and an IR scanner will cost about an extra thousand dollars. Add to this the cost of training, which should run about $450–500 per man, and you have an asset of great value to your department and community.

MAKING IT HAPPEN

It is an unfortunate fact of life in law enforcement, corrections, and military circles that when people start to move up the ladder of increasing responsibility, they get separated from those very things they used to consider important. Claiming to be able to see the "big picture," they lose sight of the little things that make up the big picture. It's really not their fault. They have a host of new responsibilities that take up an ever increasing amount of time, and the higher they go the more political their job becomes. Instead of taking time to read relevant articles, examine new (and old) techniques, and discuss new innovations with their juniors, they seem to be frozen in time, and many new developments seem irrelevant and unworthy of their attention.

This mind-set is difficult to overcome, and any argument in favor of progress will be countered with phrases containing words like "budgets," "priorities," "cost cutting," and "downsizing." As one investigator trained in tactical tracking techniques puts it, "Mantracking in the United States is not a forgotten art. Rather it is a restrained and often suppressed tool shamefully placed upon a shelf to gather dust, replaced by the less effective yet more appealing tools which provide administrations throughout the land a false perception of positive progression in law enforcement . . . new is not necessarily better."

This statement is right on the money, and it sums up the attitudes of many administrations who seek to cover up their lack of progress with emphasis on new technology before they have even mastered the old. When I was serving with an elite special operations unit I was involved with research and development, and one of my projects was the acquisition of top-of-the-line sniper rifles. After a long meeting with the unit snipers I came to the conclusion that it was not a new weapon they needed but the training and skill required to master the excellent rifles and scopes they already had.

It is too easy to blame failure on technology. That argument only works when existing equipment has been totally mastered and the required results are still not attainable. How many techniques that have proved effective in the past have been shelved in favor of new but unproven or questionable technology? Anything with words like *RAM*, *megabytes*, and *ROM* seem to be acceptable to administrations and sheriffs, but words like *eyes*, *ears*, and *intuition* are passé.

It must be realized that it is the responsibility of the team to become what they want to be. It will require self-sacrifice and dedication, but the process can be greatly assisted by proper program implementation, which is usually the responsibility of the team leader or commander. He provides the fuel to keep the fire going. There are various problems, however, common to most special teams that may have to be overcome:

- Lack of money
- Lack of interest
- Lack of administrative support (uncertainty, resistance to change, reactive versus proactive attitudes)
- Lack of training time
- Lack of equipment
- Lack of facilities

All of the above are valid problems. However, in order to make your team what you want it to be, you must set realistic goals and be determined to attain them. The formula is simple:

- Leadership
- Team unity
- Determination

Remember, regardless of how valid the reasons are for establishing a tracking team or even getting several members of your department trained, if a team succumbs to negative thinking there will always be a valid reason for not receiving training, which some day may cost a life. A bunch of interested officers who are determined to make it happen will experience progress. It may not be easy, but if you are really determined, it will happen!

Chapter 11
TRAINING THE TRACKER

An effective unit training program converts unproductive time to effective training time to not only upgrade the skills of individuals but, just as importantly, upgrade the development of the team.

U.S. Army Field Manual 21-2

Training, simply defined, consists of the actions taken to prepare an individual or unit for real situations. As much as many people will dispute it, experience by itself does not and cannot replace training.

- Individuals and teams must gear training to specific mission requirements and the kind of scenarios the team is likely to encounter.
- The training program must be safe but realistic.
- It must be effective but also address time and budget constraints.
- It should not be too complex so that it can be assimilated by all the team members.
- It must be simple and focused to address specific requirements or to improve identifiable weaknesses in the team's or individual's performance.
- It must teach skills that are relevant, have proven effective in the past, and can be recalled easily when operating under the effects of stress, fear, and fatigue.
- It must be technically and tactically correct (one can become very proficient at doing the wrong thing!). It is much easier to instill proper habits from the outset than to break bad ones later.

TEAM TRAINING

Team training must ensure that team members receive the necessary instruction and that retention of learned techniques are maintained between training periods. To be certain of this, there are basically two types of training that are designed to develop proficiency within both the individual and the team in the techniques desired:

1. Realistic training
   - increases team and individual knowledge and proficiency;
   - develops realistic SOPs;
   - presents stressful, unrehearsed planning problems;
   - combines team, individual, and technical skills; and
   - builds team and individual confidence.

2. Competitive training
   - is always performed against a set standard and time;
   - increases team potential and capability; and
   - can be done by splitting the team into sides and having them compete against each other.

Proper training habits, attitudes, and standards will ensure a tactically and technically competent team. This provides the department or unit with a disciplined, knowledgeable team that is motivated toward a successful follow-up. It also will minimize injuries and prevent potential loss of lives.

Training Program Administration

The administration of training programs should accomplish the following:

- Determine tactical policy and rules of engagement.
- Ensure weapon knowledge (employment, functioning, and safety).
- Keep written and filmed training records (particularly firearms training).
- Train on all techniques and tactics.
- Upgrade tactics and team drills to deal with current and anticipated threat levels.
- Ensure that planning skills and communications are functional and tested regularly.
- Conduct follow-up training complete with appropriate immediate-action drills (IADs) using multiple teams and air assets when possible.

This program is strongly influenced and affected by the following:

- Support from superiors
- Available funds
- Available manpower and recruiting
- Physical condition of personnel to be trained
- Training time available
- Frequency of training
- Access to suitable training areas
- Mission/role of the team (e.g., local, state, national)

Training sources and resources include the following:

- Criminal justice/POST (Peace Officer Standards of Training) sponsored courses
- In-service training
- Other police or federal agencies
- Specialist schools
- Books and training manuals.
- Independent training companies and consultants

It is important to use a balanced blend of the above resources. For instance, an agency that relies strictly on in-service training risks stagnation and has no standards or evaluations against which to measure its level of proficiency. A department with a substantial budget may send people to numerous schools and employ many outside consultants but risks confusion created by too many methods and opinions. The result is a failure to establish a solid, workable plan for the team.

It is necessary to scrutinize the contents of each program to be certain that the training is relevant, current, and conducted by people who know what they are doing. Do not become dazzled by agency titles, assuming that they must know what they're doing—unfortunately, this is not always the case. Look for operational experience and thoroughly check out qualifications by whatever means available to you. Perhaps the best way is to seek out a person or team that has been trained by the individual or company concerned and ask them for their candid impressions and comments.

(Note: The author is grateful to Steve Mattoon for permission to use the above material, which has been extracted from his *SWAT Training Manual*.)

SPECIALIZED TRACKING TRAINING

Training the tracker is not an expensive

process. It does not need access to costly equipment, scarce resources, or an inordinate amount of preparation and time. But it does need training that is relevant, creative, innovative, applicable, and interesting. Interesting training that is mission oriented is not difficult to organize, especially if team members have commitment to the team, a desire for excellence, and a determination to accomplish the mission.

When planning tracking training, team leaders or commanders must bear in mind the following points:

1. Training must be carried out in the area where tracking operations are likely to take place. It is no good, for example, to practice a follow-up in the Arizona desert if you are stationed in the pine forests of Oregon. Train in and under the conditions that will familiarize the team with what it is likely to come up against on operations. This will give team members the edge over any adversary.
2. Train with the weapons and equipment that you will use on operations. This way you will become completely familiar with both the strong and weak points of your gear and, in doing so, you will be able to reduce the weak aspects and enhance the positive ones. If your gear fails you during training, change it for a better brand or modify what you have to suit your needs and purposes. Gear that fails on operations can cause you discomfort or even incapacitate you to the degree that you are unable to carry on. Work out all the bugs well in advance.

Weapons and Ammunition

Weapons must be fired under all the weather conditions and temperatures you are likely to expect, from dry, dusty conditions to monsoonlike weather, from cold, freezing days to tropical heat waves. It is like having the ability to drive your car in all types of weather.

Heat and cold can and does have an effect on weapons. In Rhodesia, where the noon temperature often exceeded 110°F, tests showed that a cartridge that had been kept in the chamber of a rifle all day and heated up by the sun had a 12-inch difference in the point of aim from one that had been kept cool in a magazine pouch. It was believed that this factor was responsible for trained soldiers missing the first vital shots in fleeting firefights with guerrillas. As a result, certain units took the precaution of replacing this round on an hourly basis.

A word of warning: if you use one brand or type of ammunition and decide to change, make sure your weapon consistently feeds and ejects the new rounds without malfunction. There is always a tendency to train with a cheap brand (or even worse, with surplus) and go into action with the good stuff. This can be fatal unless you have completely tested the weapon/ammunition combination in advance and are happy with it.

Personnel Dynamics and Relationships

Train with the men you expect to be with on operations. Get to know each other's strong and not so strong points, and do your best to enhance the good and diminish the bad. Develop the closeness that will enable you to work together as a coordinated team with a minimum of orders. Nothing is worse than going into action with a new member whose abilities are as yet unknown. One only has to remember the "new guy" syndrome that was prominent during the Viet Nam conflict. It was always the new guy, the replacement, who was getting hurt or killed before his squad even knew his name simply because he did not know how to fit in.

Individual Team Tasks

Trackers should rotate through all the team positions, from controller to tracker to flank tracker, so that they are proficient with the specific requirements of each. The more they know about each function the more they will be of value to the team. It is inevitable, however, that certain people will become more proficient and wish to remain in one particular role. This is fine on operations because you will want the right man on the job in the place where he performs best, but it still remains important to rotate during training.

Individual Skills

Training must be planned so that each of the necessary skills for tracking are practiced in a regular and sequential manner, with each subject allocated time according to its priority. Tracking technique itself should have the highest priority, with shooting, IADs, and navigation commanding slightly less attention.

Concurrent Activity

Tracking skills should be practiced in conjunction with other applicable skills as much as possible so that trackers learn to coordinate their actions and thoughts automatically. For example, it is wise to integrate tracking training with other aspects of an operation such as formations and IADs.

TRAINING SUBJECTS

The following lists are not all-inclusive, but they contain the key elements required for training a tracking team:

1. General training
   - Tracking roles and specific techniques
   - Visual acuity and observation
   - Weapon use and employment
   - Team tactics
   - IADs
   - Navigation and GPS
   - Camouflage and concealment
   - Radio communications and SITREPs
   - Special technical equipment (e.g., heat sensors)
   - Close-quarter battle drills
   - Pyrotechnics and munitions
   - Command and control
   - Observation and surveillance

2. Individual skills
   - Proficiency at individual tracking techniques in all team positions
   - Individual tactical movement
   - Proficiency in all team weapons
   - Map reading and navigation
   - Personal camouflage and concealment
   - Radio communications and message passing
   - Use of the GPS
   - Silent signals
   - Physical fitness and endurance training
   - First aid/first responder

3. Team skills
   - Tracking formations
   - Follow-up skills (working with support elements)
   - Command and control procedures
   - Coordinated weapon handling (reflexive, direct, and covering fire).
   - IADs
   - Helicopter drills (if applicable)
   - Casualty evacuation

4. Discretionary technical skills training
   - Air operations
   - Sniper training
   - Climbing and rappelling
   - Distraction devices and tactics

These lists are by no means complete. SRTs and those teams belonging to special facilities (e.g., nuclear power stations) have mission and equipment needs not common to other teams. The above lists do, however, give the supervisor, team leader, and team members an idea of the scope of the training needs to maintain a proficient tracking team.

When assessing training needs, team leaders and supervisors must formulate priorities using the labels MUST DO, SHOULD DO, WOULD LIKE TO DO, especially if time or finance is restricted.

TRAINING SPECIFICS

The only real way to practice tracking is to get out and do it on the ground. Set various exercises for your team, from single to multiple tracks laid over different types of terrain and ground types typical to your area of operations. Select both uphill and downhill routes of varying lengths. Have the spoor layers ambush the team with slingshots—this makes them very aware of the potential of ambush and will considerably sharpen eyesight and hearing as well as hone their survival instincts.

On top of general live tracking training, a complete training program should encompass the following specific areas.

Visual Acuity

Simply put, visual acuity training entails sharpening your eyesight for the requirements of the tracking role. Most members of Western society have eyesight that has degenerated to the point where they look but they do not "see." They merely get the general picture and miss out on the details. Fortunately, most police officers have a trained eye, but if a person is taken out of his normal environment it takes time to adjust.

Trackers, when engaged on a follow-up, should be constantly evaluating and updating the tactical picture in accordance with what their eyes, ears, nose, and senses are telling them so that they can put themselves into the mind of their quarry and counter his every move. There are several exercises that can improve ones ability to "see" considerably, but remember, it is not only important to see but to interpret what we see and recall later exactly what we have seen.

Visual Acuity Exercise 1

This exercise, a mobile version of "Kim's Game," is good for all-round observation, memory, and interpretive skills.

Select a path through a wood or forest about 300 feet in length. Place along the route, in the open or partially concealed, about 10 or 12 articles relevant to a typical follow-up. These can include a backpack, shoe, cigarette carton, weapon, fired cartridge case, clothing, bloody bandage, and other items, small or large. Place them along the trail in some sort of logical sequence in an attempt to tell a story. Items must be placed at various distances from the path, from right underfoot to a reasonable distance away and at heights varying from ground level to above head height.

Get your trackers to pass slowly along the trail without stopping, visually seeking the articles and recording them to memory. When they get to the end, ask them to write down the items they saw and in the sequence in which they were laid. When the entire team has been through, have them try to deduct what has happened, first from the items selected and second from the sequence in which they were laid. Several days later, ask them to recall the exercise and repeat exactly what the articles were.

Visual Acuity Exercise 2

On a clear moonlit night, get the team to lie on their backs and "star gaze." Pick out a constellation and concentrate on it for several minutes. This will give the eyes strength and improve their ability to focus at longer distances. It is amazing that after only several minutes of this exercise, you will be able to pick out features and stars that you did not spot when you first looked. To give you an example, did you know that the "belt" on the Orion constellation has what appears to be a "dagger" tucked into it? Check it out and see for yourself. Even better, search it out and count the individual stars that make up the dagger. The features on the surface of the moon can also be used as focal points; when you have done this exercise several times the moon will look a whole lot different.

Visual Acuity Exercise 3

Lay a set of tracks through a wood or forest for several hundred yards. Return to the start point and retrace your steps. At several places along the trail, lay out several concealed trip wires connected to a military training booby trap device. Lay these out as cunningly as possible. If you are able to get your hands on some pressure release switches from an army buddy or surplus store, these can add a little spice to the exercise. The object is for the trackers to pass along the trail and see whether they can spot the booby traps before they activate them.

Visual Acuity Exercise 4

This simple exercise can be done at any time, even when you are out for a run or walk. It can really sharpen up the eyesight, especially from a tracker's perspective.

Walking or running at your normal pace, scan the ground about 20 yards ahead of you. Let your eyes pick out any abnormal objects on the ground such as cigarette butts, candy wrappers, and even cracks or bumps in the sidewalk. Quickly let your brain identify them and scan ahead to the next item. This becomes quite difficult when passing over heavily littered ground because of the mass of items that you see. Because we can see a thousand times faster than we can think, this is quite normal. If you repeat this exercise regularly, you will speed up the identification process considerably, and when you are involved on a follow-up you will be amazed just how easy it has become for your eyes to pick out the spoor.

Eye-strengthening and observation exercises are limited only by your imagination, but the important thing for a tracker is to try to involve a degree of analysis and evaluation in the game that answers the questions of what, why, who, where, when, and how.

Shooting Exercises

Each member of a tracking team must be completely and thoroughly trained in both his primary and backup weapons and weapons of his teammates. Tracking operations require a variety of weapons able to engage targets from point-blank range out to 400 yards. In desert country where visibility is good, trackers may not get the opportunity to close with and surprise their quarry, so a weapon with good long-range accuracy and hit potential is mandatory. Similarly, should the team be ambushed, it is wise to have a weapon that can fire into cover and be effective at close ranges. Close encounters of a dangerous kind are the norm for trackers, so the capability to produce short-range suppressive fire is a decided asset. Shotguns or carbine-style weapons with a quick-firing semiautomatic capability usually fit the bill.

The types and calibers of weapons selected for a tracking team will be determined by the terrain and vegetation conditions in the areas where the team will carry out its operations. Again, as a general guide, open country suggests a longer range capability and close country requires weapons with less range but increased firepower to overcome the deflecting effect caused by firing through vegetation. If, however, you are employed by a department with limited funds, you will have to become proficient with the weapons at your disposal, whether they are ideal or not for the conditions in which they will be employed.

When planning firearm training sessions, team leaders or supervisors must try to envisage all of the possible conditions and problems in which the team may be involved. Therefore, periods of training must be allocated for the following:

- Short-range engagements (up to 50 feet)
- Long-range targets (out to 400 yards)
- Moving targets
- Shooting uphill and downhill (this is always neglected and needs special attention to become proficient)
- Shooting through cover
- Shooting on the move
- Covering fire while the team maneuvers
- Poor light shooting (including in the rain)
- Cover shoots
- Jungle lanes
- Firing from all positions (standing, kneeling, lying)
- Stoppage drills and magazine changing.

Do not neglect stoppage drills and magazine changing; these are vital aspects of weapon handling and can be critical in action. The ability to clear a jammed weapon or to quickly reload and continue firing could mean the difference between life and death in a shooting confrontation.

Targets

Surprisingly, targets are a sadly neglected and misunderstood item in most police departments. One of the main reasons is because city fathers shudder in horror at the thought of their officers actually shooting at "human likenesses." As a result of this foolish, shortsighted, and ignorant attitude, departments select politically correct target types that are not at all suitable for combat training, whether it be for street officers or SWAT teams.

It is my firm conviction that many police officers who've been forced by circumstances to use their firearms in the line of duty have missed their human target because of the training they received. Paradoxically, the more training they get, the more the potential they have for missing the target. The reason is simple—many police ranges use a flat, two-dimensional target that is more often than not engaged from the frontal position due to qualification requirements and the restrictions of the stall design of the ranges.

But is this realistic for confrontations out on the street? Do criminals stand face on and shoot it out with the officer? Of course not. In most cases it is a confused, stressful, mobile confrontation, usually in poor light, shooting at fast-moving, ducking, and weaving targets. It is virtually *never* a full frontal confrontation.

Now consider the typical flat, thin, two-dimensional target with a 10X ring situated in the center of mass. The officer facing square onto the target can clearly see the six-inch ring, but if he moves to one side in the same plane, the ring appears to get smaller. As his angle to the target increases out to 90 degrees he cannot see the ring at all.

A human body has mass and depth, with the vital organs contained in the body cavity. To incapacitate a human, a bullet must be fired into the body mass, which in a real human target is up to six inches *behind* the aiming point on the flat target that he trains on. Since we tend to do in action what we are trained to do in practice, it stands to reason that the officer will instinctively select a point of aim situated on the *front* of the chest area and not on the *side* of the body where it will produce an effective hit.

To overcome this problem, the simple answer is to stop using flat, two-dimensional targets and start using three-dimensional ones, which fortunately are becoming increasingly available.

In the early 1980s, the author was responsible for the training of an elite special operations unit, and it was quickly recognized that conventional flat targets had serious shortcomings, especially when used in realistic scenarios involving fire and movement. Taking a fresh look at our needs, we came up with a three-dimensional mannequin target that fell down when struck in the critical head or chest areas. The new target system fulfilled all our expectations, mainly because to score an effective hit, the shooter had to alter his point of aim into the mass of the target where, on a real person, the heart and lungs would be located. The new target could also be engaged from all angles, which you cannot do with a flat target. It was very realistic and gave a feeling of satisfaction when the mannequin fell as a result of a correctly placed shot.

With a little imagination, everyday household items such as gallon milk jugs, plastic two-liter soda bottles, balloons, beer cans, and polystyrene blocks can be used to fabricate durable targets. Three-dimensional targets are far more realistic when they are dressed up in old clothes or uniforms and rubber masks to add variety and realism to training sessions. If your department is "financially advantaged," it is suggested you purchase the excellent three-dimensional mannequin Tri Tac target from Specialized Target Systems of Poughkeepsie, New York.

Finally, remember that combat situations are rarely static, especially after a successful follow-up when the fugitives choose to fight or flee. It follows then that trackers should be trained to use their weapons effectively against moving targets—but only if their specific rules of engagement allow! There are several excellent techniques that can be used to simulate movement, all of which are simple and cheap to set up.

During my military service my staff set up an ambush training area using a cable suspended between two trees. Standard military-type targets were suspended on pulleys, joined together by wire at varying distances, and connected to a second cable pulled by an old jeep. The targets were placed out of sight of the trainees and when the time was right (both day and night), the vehicle (with its brave driver!) would pull the targets through the ambush site, whereupon the ambush would be sprung. It was very realistic, especially because the driver was briefed to increase speed when the first shots were fired to simulate running targets.

Another excellent method to simulate running targets is to get a large car tire and cut a target so that it fits tightly inside from edge to edge. Using sloping ground, the tire is rolled across the front of the shooters, who engage the moving target. Interest can be added by rolling targets from different angles and places, causing the teams to allocate zones so that all targets are fired on.

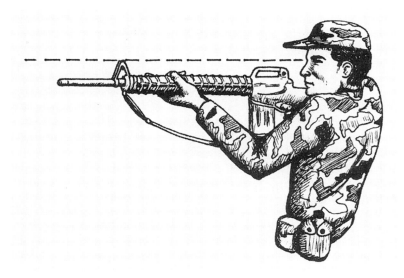

Figure 19: Quick-kill shooting. The key to successful quick-kill shooting is to keep both eyes open while looking over the sights and to align the barrel on the target. There is a natural tendency to shoot high, so a conscious attempt must be made to keep the barrel down slightly more than your instinct tells you is necessary. Two shots should be fired at the target in rapid succession, but watch for the impact point so that a fast realignment can be made in the event of a miss, when two more shots must be fired. Regular practice is the only way to master this technique.

Quick Kill

The "quick-kill" method of combat shooting, also known as "quick fire" or "instinctive fire," is an excellent and accurate technique to engage surprise targets at ranges of less than 15 yards. With practice, a good shooter can regularly hit a man-sized target without using the sights in less than a second. Both semi- and full-automatic fire modes can be used, but full-auto fire leads to a waste of ammunition, which a lightly armed tracker cannot afford to do. It is better to practice and become proficient on semiauto than to waste ammo on full auto.

The key to this technique is to shoot with both eyes open while looking over the sights to align the weapon onto the target (see Figure 19).

This has several advantages over conventional shooting techniques. First, you do not lose 50 percent of your vision by closing one eye to aim; second, you retain peripheral vision; and third, you are in a position to immediately fire again or to engage multiple targets should they appear.

Successful quick-kill shooting requires practice. This can be done by dry-firing using a snap-cap in your weapon. Practicing in front of a mirror is an excellent way to perfect the technique. (So is the use of a video camera, which, incidentally, should always be present at any firearms training sessions to record individual standards of training and proficiency. It could be invaluable later should you have to explain your use of deadly force to a judge and jury.) After making sure your weapon is empty, stand in front of a full-length mirror (not too close or you may strike the glass) with your weapon in the normal patrol carry. At a given signal, bring the weapon up to your shoulder and, looking over the front sight, quickly align the barrel

Training the Tracker

Figure 20: Pointed quick-kill shooting. Pointed quick-kill shooting is generally used to engage surprise targets at ranges of less the 10 yards. The weapon is not brought to the shoulder as in quick kill but aligned on the target from the waist. Two rapid shots are fired; if they miss it will take only a fraction of a second to realign and fire again. The barrel is not aligned in front of the eyes, so the difference in angle will have to be compensated for.

onto your reflection and squeeze the trigger. You will clearly see if the weapon is aligned correctly with your center of mass, which should be the upper chest area at approximately the center of a cross formed by a horizontal line across the armpits and a vertical line through the sternum. Initially you may tend to aim a little high, so concentrate on keeping the barrel down. After repeating this exercise until you are certain you can align your weapon correctly every time, your muscle memory will be imprinted with the correct "feel" of the position and you can now graduate to live firing.

Another useful technique is one the military calls the "pointed quick kill." It is used to engage surprise targets at ranges of under 10 yards. When presented with a target, the weapon is not brought up to the shoulder in the usual way but is aligned on the target from the waist (see Figure 20). Line up the barrel and squeeze off two rapid shots. If you miss the target it will only take a fraction of a second to realign and fire. Pointed quick kill is a little faster than quick kill but is not as accurate because the barrel is not aligned on the target in front of your eyes and the difference in angle has to be compensated for. However, as with everything else, you will improve with practice.

Practice, practice, practice is the key to quick-kill shooting. It must be stressed that it is only used to engage surprise, close-range targets and is *not* a substitute for a well-aimed shot using a proper sight picture.

Jungle Lanes

An excellent training method for trackers is the "jungle lane," which entails movement, observation, recognition, and shooting skills—all those things that a tracker must be proficient at during a follow-up.

The jungle lane consists of a trail, about 200 yards in length, running through an area in the type of terrain where you would expect to be involved on operations. A number of targets—some static, some movable, and some pop-up—are placed along the trail. The tracker is given a period of time to move along the lane, observing carefully and engaging the targets he sees with two quick-aimed shots. The use of reactive or fall-down targets adds realism and stress as well as ensuring that the shooter engages the targets correctly, because he is not allowed to proceed until the target is down. Limiting the amount of ammunition used on this exercise requires the tracker to shoot effectively the first time.

This shooting exercise can be done individually or as a team and is as close to the real thing as any training scenario can be. Objects such as bloody bandages, blood drops, weapons, and other items of equipment can be

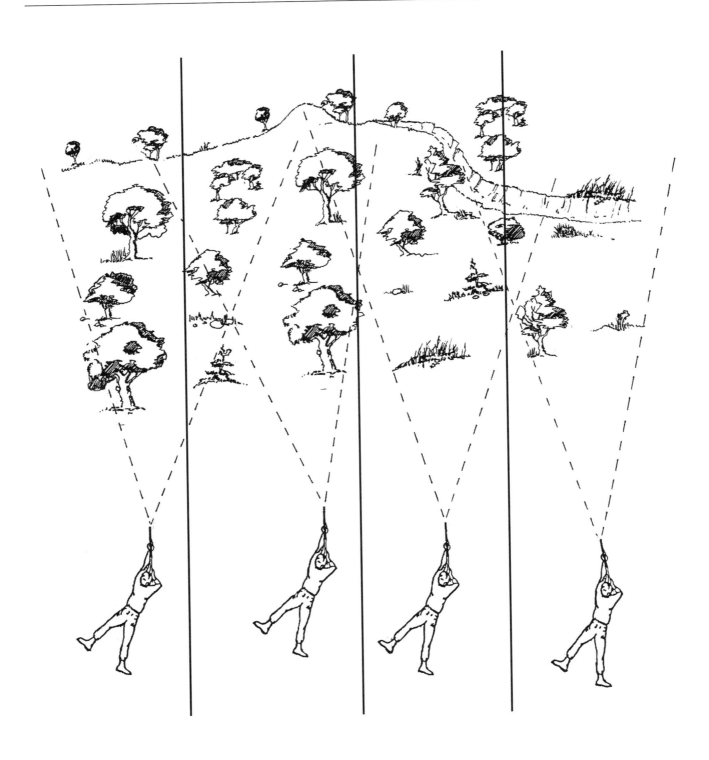

Figure 21: The cover shoot. Each tracker fires two aimed shots into positions of likely cover within the zone immediately to his front. Starting close up, he systematically progresses further and further back, widening his arc of fire, until all likely positions have been engaged. Shots are placed into each side of the cover at close to ground level. This technique is effective in flushing hidden adversaries and is economical in ammunition expenditure.

placed along the trail. Booby traps to catch the unwary add a lot of interest to this training.

If run on a competitive basis for individuals, scoring takes into consideration targets hit correctly, amount of rounds fired, and recall of objects placed along the trail. Penalties can include the amount of targets and trip wires missed and booby traps activated. Jungle lanes are good training and good fun, but they do take time to set up, and sufficient time must be allocated for everyone to have his turn.

Camouflage and Concealment

The aim of camouflage and concealment is to confuse an observer into believing he is seeing something other than what he thinks he is seeing by distorting and altering its physical characteristics. Having said that, camouflage is no good on its own; a perfectly camouflaged team will stand out like ball on a snooker table if it ignores other aspects of proper concealment. Some things that need to be avoided include these:

- Wrong setting and background
- Unnecessary or rapid movement
- Noise
- A UV signature
- Casting shadows
- Bunching up

To repeat what was said in Chapter 1, trackers have to follow the tracks wherever they may lead, and this can lead them into grave danger if the quarry knows he is being followed and prepares an ambush. There are several ways a tracker can reduce the chances of being seen prematurely and minimize this tactical disadvantage:

- Move as quietly as possible.
- Minimize all radio traffic or, better still, switch off all radios.
- Use shadow, cover, and ground to advantage, especially the controller and flank trackers.
- Maintain a loose formation and don't bunch up.
- Adopt an aggressive camouflage policy.

All of the above are important, but probably the most effective is an aggressive camouflage policy. To achieve the tactical advantage that effective camouflage provides, the team must pay maximum attention to detail. Clothing, weapons, skin, and footwear must all be minutely scrutinized and attended to. Remember the points that need to be addressed when applying camouflage:

- Shape
- Shine
- Shadow
- Silhouette
- Surface
- Spacing
- Movement
- Noise
- UV radiation

When I was a young soldier in Rhodesia, I recall our first lecture on camouflage. Three of our recruit squad were pulled out for a demonstration, and black camo paint was applied to their faces. The first was "Too Little"; he had only a few smears of paint randomly applied to his face, leaving the highlights untouched. The second was "Too Much," his face being liberally smeared with paint, Al Jolson style, until no skin was showing. The third man was "Just Right," with paint correctly applied to neutralize his facial high points so that the face lost its familiar shape and individual characteristics. The three soldiers were then placed into cover and we could clearly see the point of the exercise. Too Much and Too Little stood out, and Just Right blended in with the background. The point was driven home and I have never forgotten it.

The best way to go about mastering effective camouflaging technique is by trial and error in training. Take the team into the woods or to the type of terrain and vegetation in which you expect to operate and play around with different ideas. Using paint, camo cream, scrim, netting, burlap, and even plastic leaves obtained from a craft store, experiment until the best effect is obtained. Test your designs by placing individuals into various types of cover and see whether the rest of the team can spot them.

This exercise does not require going to the extent of making or purchasing an expensive ghillie suit, which are all the rage in the military catalogs these days. Remember that ghillie suits are hot to wear on a follow-up, and they considerably reduce your ability to move, hear, and see effectively. They're fine for sniping or reconnaissance work, but they have no place on a tracker. Try to stay as simple as possible and fit into your environment without sacrificing freedom of movement or impairing your sight and hearing. Simple yet effective is the key.

The camouflaging of weapons can present problems, as it is unlikely that a training officer is going to permit a department weapon to be painted brown and green. When I painted my Rhodesian Army FN FAL, the first reaction was shock and horror on the part of my superiors. There were even murmurings of court martial! Fortunately common sense prevailed, and it was not long before a general order was issued to camouflage all operational weapons. However, in cases where department policy or, even worse, senior commanders' prejudice and ignorance prevents you from permanently painting your weapon, experiment with cloth covers that can both be removed and do not interfere with the functioning of the weapon. Camouflaged gun socks or adhesive tape are quite satisfactory and can be obtained from most sporting goods stores. Even better is to have the weapon treated with a flat, matte, durable camouflage finish.

A North Dakota gunsmith, Rick Wagner of Wagner Companies, offers an excellent camouflaging service called Forever Blue. He applies a high-tech film coating to the external surface of the weapon that is corrosion resistant. The result is a worry-free camo finish that will last for years without attention. Finishes available include the plain colors such as green or brown or any pattern of colors, depending on your requirements.

Rick also offers an interior finish for firearms called Forever Lube that bonds a dry-film, high-tech lubricant on to all internal surfaces, including the bore and working parts. The film provides long-lasting protection against corrosion and never needs lubrication. This treatment increases the life of the weapon, and working components stay free of carbon buildup, thus ensuring greater reliability. Even a neglected, rusty weapon with a pitted bore or one damaged in a fire can be treated by this process, giving it a new lease of life. I have a 50-year-old M1 carbine treated by Rick that is totally maintenance free. His prices are reasonable, and his turnaround time is good.

Whichever method you use to camo up your weapons, bear in mind that it must be done with as much care and attention to detail as you would give to your own appearance. When you are tracking, your weapon is held in front of you in the ready position, and if it is not camouflaged the light reflected off the barrel, magazine, or other shiny surfaces will inevitably be spotted by your quarry. Reflected light can be seen 10 miles away if conditions are right. Don't forget to treat all of your spare magazines too.

Many millions of dollars have been spent over the years researching different methods of camouflaging men and material. The reason for this is simple—if you can see a person or thing, you can shoot at it too. If your enemy cannot see you, he cannot shoot you, so the aim of the game is to make it as difficult as possible for him to line you up in his sights.

So seriously does the U.S. Army treat this aspect of soldier survival that a research team at the army's Natick Research and Development Center is investigating how uniforms can be made to change color instantly and automatically depending on the color of the surroundings. Known as "adaptive coloration," teams of scientists are studying the way chameleons are able to change color as well as experimenting with electrically charged colored particles incorporated into flat fibers that can be woven into a color-sensitive fabric. It will be several more years before a cost-effective technique is developed, but military authorities are already hard at work revolutionizing the high-tech, computerized battlefield of the 21st century. In the meantime, we have to "soldier on" with what is available to us.

During my service with the South West Africa Territorial Force in what is now Namibia, I personally inspected all the patrols leaving my base, which was situated only a few miles from the hostile Angola border. My instructions were for all troops leaving base to adhere to simple but effective camouflage requirements. This included the use of green and brown nylon scrim on their plain brown boonie hats, which we trimmed from the camouflage netting that covered the perimeter bunkers. Apparently this Ramboesque policy was greeted with a lot of snickering and scorn by officers and troops of other units, who obviously could not see the point of doing things correctly.

One evening, a patrol had moved to within a few hundred yards of its intended ambush site, a path crossing the international border with Angola. The platoon commander, after ordering his 20 men to remain in a position of all-round defense, moved forward to reconnoiter the ambush area. Several minutes later the watchful patrol, crouching down in the knee-high grass and scrubby bushes, heard voices from the direction of the border. Looking in that direction, they were surprised to see five heavily armed guerrillas of the Peoples Liberation Army of Namibia moving directly toward them, totally oblivious to their presence. I am sure my troops held their fire out of sheer fear, but the hapless guerrillas, still oblivious to their impending doom, walked up to within 10 yards of the troops, who at last did the right thing and opened fire with everything they had.

My temporary tactical base was only about a kilometer away, so I moved up to the contact site quickly to investigate. I could clearly see exactly where the troops had been positioned by the heaps of fired bullet cases, and the closest guerrilla corpse was no more than eight paces from the line. The amazing thing was that there was very little cover in the area; just the occasional tree and scattered bushes about four feet high. The success of the improvised ambush proved beyond doubt that even the simple steps we took to break up the distinctive outline of the bush hats, combined with the troops' black faces, paid handsome dividends.

When we returned to base several days later, something strange happened. Everybody, including the erstwhile skeptics, were sporting camouflage on their hats and equipment. How odd, I thought, I wonder why?

Team Tactics and Immediate-Action Drills

Team tactics and IADs were covered in Chapter 5, so I will not repeat them here. Suffice it to say that this critical aspect of tracking operations must be practiced again and again to achieve reactive proficiency. Tactics and IADs must be completely understood by the entire team so that when contact is made with an armed quarry, every man instinctively knows exactly where he should go and what he must do when he gets there. To attain a high level of teamwork means working through as many scenarios as possible. Probably the best way to achieve this is to ask cooperative members of your local National Guard unit to assist you by providing several willing volunteers to act as spoor layers and fugitives. With their military training they will have a good idea of how to make it interesting for you and provide many different scenarios and encounter actions for you to respond to. (Don't forget to buy them a beer afterward.)

It is important when practicing these drills to take time out after each scenario to conduct an after-action critique. You will learn from your mistakes, if any, and it also encourages newer or younger members of the team to contribute their ideas and suggestions. Not only should you examine and analyze your actions, you should also discuss what other methods you could have used to achieve your objectives. Do not fall into the trap of discussing the thing to death, and remember that there is no one "right" way to achieve a tactical aim; there may be several ways to accomplish the same purpose. The more flexible you become in your approach, the better you will be able to cope with more complex tactical situations that *will* occur when you are up against a real foe.

While you are engaged in tactical training, try to add an element of realism or surprise by including a simulated casualty or the addition of strategically placed booby traps. It is as well to remember that realistic, creative training will go a long way to improve the versatility and capabilities of your team as well as give you a lot of fun.

Radio Communications

No special requirements are necessary to practice communications. In the course of their normal duties most police officers are on the radio continually. There are two points to emphasize, however, in respect to the tracking role: keep radio transmissions as brief but as complete as possible, and resist the urge to use the radio unnecessarily. Nothing is worse than being cut out of the net when you have an immediate tactical message to pass because some idiot is prattling on with nontactical nonsense. You've all heard 'em!

If it is known that your opposition is in the habit of monitoring your radio transmissions (there are a lot of police scanners out there), it is well to devise and use some form of simple code. If you feel that this is advisable, schedule practice sessions until everyone involved is familiar with sending and receiving encoded messages. Remember the military acronym SAD—security, accuracy, and discipline.

Radio transmissions can be vulnerable to jamming or interception and can be located with direction-finding equipment. Steps should be taken to devise alternative frequencies and call signs known to the whole team if you encounter such circumstances.

Navigation and Map Reading

In the section on the global positioning system, it was mentioned that technical aids can never replace the ability to read a map and navigate over unfamiliar terrain. Batteries fail, especially in cold weather, and most high-tech toys develop glitches from time to time. There are many good books and videos available on the subject of map reading and navigation, so it is not my intention to cover the subject in this book. If you are unsure of your map-reading skills, do yourself a favor and visit your local book or sporting goods store and examine the offerings. There is one for every pocketbook, from several dollars up to more expensive, degree-level versions.

Map reading is not difficult. All that is required is the ability to "see" a map as a three-dimensional aerial picture of a piece of ground. The trick is to relate this picture to the land forms you see around you, recognize the man-made items such as roads, railway lines, and buildings, and imagine yourself moving, antlike, across the paper. As you proceed, seek out and identify upcoming hills, rivers, and man-made features on the ground and relate them to the markings on the map. Once you have grasped this simple principle, map reading and navigation should be a lot easier to understand. The key, of course, is practice, practice, and more practice. The strange thing is, the more you practice the better you get at it.

For those people who really enjoy navigating in the wilds, there are orienteering clubs all over the country where teams and individuals compete against each other on a prearranged course from checkpoint to checkpoint in the fastest time possible. Clubs are willing to teach newcomers and encourage participation in their competitive events.

Any police or corrections officer working in a rural or remote area really should have a good grasp of navigational techniques, especially if he is expected to venture off the blacktop and into the woods. It is not good enough to rely on local knowledge, especially if you have to give directions over the radio. What sounds better: "From the big hill with the white rock on top, go south three miles until you get to a small stream, take a left, and you will see a clump of rocks about half a mile to the east. We are at the base of the biggest rock," or, "Proceed to the rocky outcrop at grid reference 752817." I know what sounds better to me.

As far as tracking is concerned, the ability to read a map accurately is of the utmost importance, especially if you are engaged on a follow-up of dangerous fugitives. As well as your life depending on it, your command and control officers are going to want to know exactly where you are and, even more important, where you are headed so that they can deploy other assets to cut off or ambush the fugitives. If trackers on the spoor pass wrong grid references to control, the follow-up and indeed the whole operation may be placed in jeopardy. In a worst-case scenario the fugitives may escape to prey on society again.

During the Rhodesian war, one of the best trackers we had miscalculated his location by 10 kilometers (nearly six miles). This placed him in a free-fire zone and tragically led to his untimely death, shot in poor light by one of his best friends serving in another unit. Wrong locstats can ruin a follow-up or even get you killed.

Command and Control

As with any other complex military or police operation, the command and control team responsible for tracking operations requires regular rehearsals to achieve and maintain proficiency. (See Chapter 8, Command and Control of a Tracking Team.) It must be remembered, however, that in the case of a breakout by dangerous prisoners from a remote jail, for example, the tracking element will be only a small part of a much larger operation involving many other law enforcement and corrections personnel.

It is not necessary to involve all the different elements that make up a total operational HQ to practice tracking command and control. All that is required is a commander, radios, a map board, and one or more tracking teams to simulate a realistic and worthwhile exercise.

Using anyone available—National Guard troops, off-duty buddies, or compliant family members—to play the role of the bad guys and lay the tracks, the tracking team is transported by whatever means available to the follow-up commencement point. From there they go through all the necessary activities required to establish their opening SITREP of LNDAT (location, number of tracks, direction, age, type of tracks). The exercise then continues in the same way as a regular real-life follow-up. If additional assets are not available, the commander in the operations center simply passes fictional messages to the team on the tracks to simulate additional activity. For example, he could place observation posts or spoor cutting teams on the map and inform the follow-up team of their positions. In this way a complete follow-up can be rehearsed without the need of an entire command and control staff.

Another method of exercising command and control staff is to use a radio to transmit messages from a prewritten script into the operations center. By condensing time, an entire follow-up can be conducted within a couple of hours, far shorter than the time it would take to conduct a real-life operation. (Such an exercise is outlined in detail in Chapter 8.)

Innovative and creative training *must* be done to iron out in advance unsatisfactory communications, poor liaison, lack of transport, and other potential problems. This type of training does not have to be done on a regular basis, but anybody who can expect to be involved in a command and control position in a live follow-up must be aware of what is required, what is important, and how to solve problems.

Observation and Surveillance

We have already established the fact that due to their intimate knowledge of rural operations and fieldcraft skills, trained trackers make excellent observers. Like everything else, observation and surveillance skills have to be rehearsed, revised, and honed to the point of near perfection to be of any value to tactical operations. The greatest sin for a recon team is to be compromised during the surveillance phase of an operation. If the perpetrators are warned that an operation is pending, they will go to great pains to conceal, suspend, or move their illegal activities elsewhere.

To be compromised during surveillance means that the operation will have to be terminated, possibly throwing away months, even years, of painstaking investigation and research. On the other hand, there are times when surveillance operations have been compromised through no fault of the observers. Bad luck and pure chance have often played an unwelcome role in operational failure. Should compromise or even the suspicion of compromise occur, there is nothing else to do but go back to the drawing board and start over. Thinking the perps will continue to do what they were doing knowing they are being watched is foolish in the extreme. Sometimes hard decisions have to be made, and this is one of those times. The decision to abort observation *must* be made by the observers themselves and not by command and control staff members. The man on the ground is the only one who knows the full facts of the situation, and the sole responsibility should rest with him. The command staff, safe and warm in its HQ, should never micromanage operations or second-guess the men in the field. To do so shows an extreme lack of command and control experience and zero common sense.

Surveillance and observation training should be conducted in the same way as a real operation. Orders must be given, and the recon team must have the opportunity to plan and conduct the mission itself. The subject of the exercise could be anything from watching the activities of a farmer going about his normal daily activities to a prearranged sequence of events taking place over a period of time using officers or co-opted friends playing the role of the bad guys. The recon team will be required to keep accurate logs of activities and incidents that can be compared with the prepared timetable at the brief-back session. All aspects of infiltration and exfiltration should be covered, as should simulated problems such as injuries, compromise, obstacle crossings, and equipment failure. Inclement weather should never be considered a factor in any tactical training exercise despite any real or imagined discomfort felt by the participants.

Essential equipment required for observation and surveillance tasks include the following:

1. Day tasks
   - Binoculars
   - Tree-climbing equipment
   - Camera, still or video
   - Tape recorder

2. Night tasks
   - Starlight scope or other weapon-mounted night-vision device
   - Infrared devices
   - Camera with high ASA film (3200 or 6400 ASA for night use)
   - Night-vision goggles

With a little creativity, a realistic exercise can be conducted with a minimum of personnel and props. If a team is engaged in real-life observation and surveillance tasks on a regular basis, it is not necessary to carry out a program of ongoing training. However, it is wise to conduct one such training exercise at least every six months to ensure that bad habits have not crept in that so very often become established as the norm. It is easier to break bad habits during training than it is to do so on operations when so many other things have priority.

SUMMARY

Make training interesting, innovative, and mission oriented. Integrate as many skills as you can when engaged on tracking exercises, and get different people each time to lay the tracks. Introduce antitracking, countertracking, ambushes, and as many wild cards as you can dream up. Arrange contests with other teams and units and discuss training experiences together. Individual and team skill levels will rise substantially and, most important of all, you will retain the interest, cooperation, and respect of your team. In this way you will create a team with the ability to accomplish any task and overcome any problem.

Chapter 12
WEAPONS AND EQUIPMENT

For the want of a nail, a shoe fell off. For the want of a shoe, a horse was lost. For the want of a horse, a knight was lost. For the want of a knight, a battle was lost. Because the battle was lost, the kingdom was lost. All for the want of a nail.

Traditional fable

No book of this type would be complete without a discourse on the weapons and equipment suitable for tracking operations, particularly the used in pursuit of dangerous fugitives. All of the weapons, ammunition, and equipment discussed here have been personally tested by me and are readily available in the United States. Chances are that you will already have a large proportion of them in your inventory.

I must state at this point that although certain brand names and models are mentioned, at no time have I agreed to sponsor or promote any individual manufacturer's product or products. By the same token, I disavow any responsibility or liability that may occur from the use or misuse of any of the products mentioned in this book. They are mentioned simply because I consider them to be eminently suitable for a tracking operation due to their reliability, efficiency, and ease of acquisition. There are other excellent products on the market that are probably as good as the ones I have chosen to describe, but they are not discussed simply because I have not used or tested them myself.

I do not discuss handguns in this book. Some models and calibers are more suitable than others, but because most trackers from both police and corrections departments have to conform to departmental policies, it would be impractical to recommend weapons that would not be permitted anyway. Also, handguns are secondary to long guns in the tracking role and would only be used as a last resort. In more than 20 years of active military tracking operations, I never used my Browning 9mm pistol once, although it never left my side.

AMMUNITION

Irrespective of the type or caliber of your weapon, always ensure that all the ammunition you bring on a follow-up is of the best manufacture and in good, fresh, clean condition. A good way to ensure this is to use this "first-line" ammo for each training session and replace it with fresh every time. This method has two benefits: you assure yourself that the ammunition you have chosen functions reliably in your weapon, and you are obliged to purchase fresh ammunition to replace what you have expended on training. Should you switch brands or types, spend time rezeroing your weapon and sort out all feed problems, if any, well in advance.

Military surplus ammo is relatively cheap and good for practice, but it contains too many unknown variables for you to risk your life on it during operations. Similarly, remanufactured ammo is great for practice but not good for ops. I recall the time I had a malfunction with some commercially remanufactured .45 ammunition. When I pulled the head to examine the case, I found it packed with well-chewed gum! *Always* stick to good quality commercially manufactured ammo when your life is on the line.

The five most common calibers available to law-enforcement agencies and military teams for long or sub guns are these:

- .223 (or 5.56mm)
- .30 (M1)
- 7.62x39mm (Soviet)
- 9x19mm (Parabellum)
- .308 Winchester (or 7.62x51mm NATO)

All of these are of military lineage and have been tried and tested in countless conflicts worldwide over many years, so there can be no doubt as to their efficiency.

Weight

As far as suitability for the tracking role, the weight factor has to be taken into consideration: a lightly equipped tracker should endeavor to strike a balance between efficiency and mobility.

Those of us who have carried large amounts of ammunition on military operations when resupply has been a problem know full well that ammunition is heavy and it can be expended extremely rapidly in a firefight. The cyclic rate of fire for some full-auto weapons can exceed 600 rounds a minute. Since weight and firepower are vital factors, the choice of caliber and the number of rounds carried are critical. Assuming that 200 rounds are considered reasonable for a tracker, let us have a look at the weights of four of the above calibers:

Caliber	Weight (grains)	Number of rounds per 100	Weight (lbs.) per lb.
9mm	129	37	5
.223	55	38	5
7.62x39	120	29	7
.308	150	19	10

(Source: G. Markham, *Guns of the Elite*)

As can be seen, from a weight point of view the .223 has the edge, being half that of the .308 Winchester. But there are other factors to consider.

Range and Velocity

Given that the maximum range required of a trackers weapon is 400 yards, let us examine effective ranges, in yards, and velocity, in feets per second (fps) of each caliber:

Caliber	Average Shooter	Expert	Average Velocity
9mm	50 yds.	100 yds.	1,100 fps
.223	250 yds.	400 yds.	3,200 fps
7.62x39	200 yds.	400 yds.	2,400 fps
.308	300 yds.	600 yds.	2,850 fps

(Source: G. Markham, *Guns of the Elite*)

Clearly, the .308 NATO round has the edge here, having a greater range and accuracy potential than the smaller, lighter .223 and 7.62 Soviet projectile.

WEAPONS SUITABLE FOR TRACKING

The three types of weapons reviewed below are the assault rifle, shotgun, and submachine gun. Each weapon and its ammunition will be examined in terms of its weight, firepower, range, ammunition capacity and weight, and handling characteristics specific to the tracking role.

As has been stated elsewhere in this book,

Weapons and Equipment

Left: The M16/AR-15 series of weapons in .223 caliber are ideally suited for tracking operations in open or sparsely vegetated country. The light, fast, 55-grain bullet, however, leaves a lot to be desired if firefights occur in thick brush: the round tends to be deflected off twigs, leaves, or grass. Better to employ heavier, harder-hitting bullets such as the 7.62x39 or 7.62x51 (.308) or even a shotgun under these conditions.

Right: The sliding-stock carbine version of the M16 is light, portable, and manueverable, making it the first choice of some police trackers. Better if it were 7.62x39mm, though.

weapons for trackers must be capable of engaging targets effectively from point-blank range out to 400 yards. This may well entail a mix of shotguns and rifles, which is acceptable and under some conditions even preferable. Again, the primary consideration when choosing a tracker's weapon is the type of terrain in which he can expect to operate. Shotguns are not very effective in the Nevada desert, where longer ranges are likely. Likewise, a .223-caliber assault rifle is not the best choice in thick brush due to the ammo's tendency to tumble and be deflected when fired through bushes or grass.

.223 Assault Rifles

There are many excellent weapons available for the tracker in the .223 caliber.

M16/AR-15 Series

The AR-15s—including the ubiquitous M16A1 and A2, the M4 carbine, and the commercial variants—are highly suitable for employment in tracking. They provide high velocity, intrinsic accuracy, firepower, ease of handling, and availability. Far more ammunition can be carried than with other calibers, and the weapon has been battle tested for more than 30 years.

Whereas the basic models of the M16, the A1 and A2, are suitable, there are several variants on the market available to law enforcement agencies that will find favor with trackers. The CAR 15, a short-barreled version of the M16, was the choice of many members of long-range reconnaissance patrols during the Viet Nam conflict as well as SWAT teams throughout the country. It has a collapsible sliding stock, making it lighter and handier than the standard M16. Currently

The short-lived Armalite AR-18. Considered by the author to be superior to the M16 series, it was not manufactured in large quantities and was not adopted by any army other than for special operations use. A favorite of the IRA in Ireland, it is light, accurate, and maneuverable, making it ideal for the tracking role The author used this weapon on tracking operations in Rhodesia and South West Africa and recommends it highly.

manufactured by Colt and Olympic Arms, several models are in production. Besides the Colt CAR 15 and M4 carbine, highly recommended is the Olympic Arms Multi-Match ML-1 version, which has great accuracy and weighs a little under six lbs.

Armalite AR-18 and 180

Unfortunately now out of production is the Eugene Stoner-designed AR-18, a weapon that I carried for some time during the Rhodesian bush war. Manufactured originally in the United States and later under license in Britain and Japan, this compact, lightweight rifle is ideal for the tracker. I did find several minor faults with it, though. Because of its shape, the oddly bent charging handle tends to catch in thick brush, and the large magazine release catch can be depressed inadvertently, dumping the magazine on the ground at embarrassing times. To my mind this weapon is an improvement on the M16, but due to the worldwide adoption of the M16 and its variants, the AR-18 has fallen by the wayside.

Steyr AUG

The Steyr AUG has been available recently to law enforcement agencies on a limited basis but unfortunately is one of the weapons forbidden in the U.S. under the so-called assault weapons ban of 1994. If you are fortunate to belong to a department that had the foresight to purchase this excellent weapon, you are in good company. Some of the world's finest military special operations units, including the British and Australian SAS, have adopted the AUG because of its superior characteristics. Its compact bullpup design and large proportion of synthetic parts make it ideal for trackers. The only drawback to my mind is the optical sight, which is not as fast to acquire as a traditional iron sight, especially for people who have trained themselves to shoot with both eyes open in combat situations.

Ruger Mini-14 Series

Probably the best of the .223-caliber weapons suitable for trackers is the Ruger Mini-14 or one of its variants. This excellent weapon is a combination of the old M14 service rifle and the U.S. M1 carbine of World War II fame. It is an exceptionally robust firearm of mechanical simplicity and excellent accuracy. Although sold with wooden furniture, synthetic versions with both full and folding stocks are available from commercial aftermarket manufacturers.

The Mini-14 and the full-auto AC-556 version have been purchased by several U.S. government agencies and military special operations units that consider it to be a first-rate counterterrorist weapon. It fires from a variety of box magazine capacities, from 10 to 50 rounds, although the 20- or 30-round versions are more efficient and reliable for normal operational use. Also made in stainless steel, the Mini-14 is the lowest priced .223 semiauto on the market today and is an ideal choice for a cash-strapped police department. Already a favorite with many corrections officers, it would be an eminently suitable choice for a corrections tracking team.

Foreign Weapons

South African weapons, once on the banned list, are slowly finding their way into the United States. One of the finest .223 military assault rifles ever made was developed from the Israeli Galil (itself an AK-47 variant) for the South African Defence Forces. Called the R4 by the military, the series is sold in several civilian semiauto models of varying barrel and body lengths. Known as the LM5, LM6, and LM7, they are ideal for a tracker. If you are able to get your hands on one you are a fortunate man indeed.

7.62x39mm Assault Rifles

The choice of a Soviet-designed weapon system may seem to be strange for a law enforcement tracking team, but there are good reasons why this caliber should be seriously considered.

AK-47 and AKM

After a number of years on the wrong end

of the AK-47, I, along with thousands of Viet Nam vets, have learned to treat the "Kalash" with the greatest respect. Designed to be used by peasant soldiers, the Soviet designer Kalashnikov came up with a weapon that just went on ticking irrespective of how much abuse or lack of maintenance it was subjected to. Several models of this ubiquitous weapon are available in the United States today at a cost far less than any domestic equivalent. Most models sold commercially have been modified to semiauto only, but select-fire versions can be legally purchased by police agencies.

The AK is a first-rate, light, accurate, portable weapon ideally suited to the tracking role. Both 30- and 40-round magazines are readily and cheaply available for it, and there is even a 75-round drum if you are looking for that "Schwarzenegger" edge.

SKS

As is the case with the AK-47, the U.S. market has been flooded with millions of Russian and Chinese versions of the semiautomatic Simonov carbine, the venerable SKS. Designed for use by a peasant militia, it is virtually abuseproof. As originally designed, the SKS fired from a fixed 10-round box magazine that could be rapidly reloaded with stripper clips inserted straight into the magazine through a slot in the upper receiver. Several aftermarket manufacturers have come up with 20-, 30-, and even 50-round detachable magazines, which add a variety of options that considerably enhance this simple and sturdy weapon. With a barrel slightly longer than that of the AK-47, it is more accurate but only fires in the semiauto mode, which some may consider a disadvantage in a firefight.

Several versions of the SKS are available, most of them modified to comply with current BATF regulations. Thumbhole, Monte Carlo, and synthetic stocks are available, as are a host of aftermarket accessories necessary to customize the rifle to your heart's desire. Ranging in price from $100 to $175, depending on whether it's of Chinese or Russian fabrication, it is certainly the cheapest rifle on the market today. (There is even a Chinese paratrooper version available. No matter that China never had any paratroops, the marketing people will come up with anything to make a buck. This short-barrel version is well made, compact, and lends itself well to customizing.) SKSs are so ridiculously cheap that one can experiment with all manner of modifications and camouflage designs. Given the number of these rifles in the country and the penchant for the BATF to ban such weapons, it can be assumed that many will end up being confiscated and available for departmental use if desired.

Ruger Mini-30

Several years ago Ruger, maker of the excellent .223 Mini-14 series, experimented with a weapon modified to fire the 7.62x51mm NATO cartridge. Despite extensive trials and experimentation, it was discovered that the weapon lacked acceptable accuracy in that particular caliber and the project was abandoned. Ruger engineers did not forget the project, however, and when Soviet bloc 7.62x39mm ammunition became available in enormous quantities in the United States, they combined the proven Mini-14 system with the Soviet M43 round and came up with a winner, now known as the Ruger Mini-30.

Despite the fact that the bullet was more than twice as heavy as the .223 (compare 55 grains to 120 grains), the mating was ideal, with a minimum of reengineering and design changes required. Having almost the same weight and handling characteristics as the Mini-14, the Mini-30 has the added advantage of firing a heavier bullet, which improves performance considerably when fired through vegetation or light cover. The steel-jacketed, steel-cored bullet (now banned for civilian use), although 800 fps slower than the .223, is less prone to deflection and tumbling, which severely downgrades the lighter, faster .223 bullet in heavily vegetated areas. Its accuracy, lightness, and reliability, combined with good firepower and superior ballistics, makes it almost a perfect choice for a tracker weapon.

Olympic Arms Hunter

In the same way that Ruger married the Soviet 7.62x39 round to the Mini-14 system so successfully, Olympic Arms of Tacoma, Washington, has done the same thing in its AR-15 Hunter, which combines the 7.62x39 round with the M16 series. Available with either a 16- or 20-inch barrel, the Hunter is a marriage made in heaven for the tracker and is probably the most ideal weapon for the role. A superior advantage of the Olympic Arms series is the ability to convert the weapons of that series to a variety of other calibers just by changing the upper receiver. There is a slew of options available: 9mm, .45, .40 S&W, 10mm, .223, and of course the 7.62x39, all using the same lower receiver. This is truly an awesome number of combinations in undoubtedly the most versatile weapon system ever offered to law enforcement.

.30 Carbines

Although I had not originally intended to include the .30 caliber in this section, recent technological advances have brought the venerable 50-year-old M1 carbine back to life and into contention as a tracker's weapon.

M1 and M2 Carbine

Invented in jail by a man serving a life sentence for murder, the M1 carbine was hastily introduced to U.S. soldiers to replace underpowered military handguns at the commencement of American involvement in World War II. Over 6 million M1/M2 carbines eventually saw service in all World War II theaters and the Korean War. In early 1996, the U.S. government shipped 30,000 surplus M1s to Mexico for use against antigovernment guerrillas operating in the southern jungle regions of the country, so even after 50 years of service it still plays a role in combat operations.

Weighing in at around five lbs., the M1 carbine is a compact, handy weapon, but it suffers badly by being underpowered. The 110-grain bullet traveling at a velocity of around 2,200 fps just was not enough to guarantee effective stopping power, so the weapon was eventually phased out of military service before the war in Viet Nam. Despite its shortcomings, its availability and low cost make it attractive to a small but enthusiastic band of followers who have a lot of fun firing available, inexpensive surplus ammunition.

Recently, several arms companies have taken a fresh look at the old M1, and updated versions have appeared on the scene. Recognizing the fact that its main weakness is its lack of punch, a metallurgist employed by General Motors Corporation, Tim LeGendre, used his knowledge of special steels to design and manufacture high-tech barrels that could handle more powerful modern magnum ammunition.

LeGendre's first successful project was an updated M1 carbine known as the MAG-1, which fires the powerful .45 Winchester Magnum pistol cartridge. For about $350, any M1 carbine can be converted to the new caliber. The barrel is made of a blend of two types of experimental steel that is strong enough to handle the high pressures generated by magnum ammunition. As well as fitting a new barrel, the bolt face and extractor are modified to fit the .45 Winchester Magnum cartridges. The original magazines are used and available in 6- and 13-round capacities. LeGendre plans to offer other chamberings, including 9mm Winchester

The venerable but highly efficient FN FAL in .308 (7.62x51 NATO). Used by more Western armies than any other weapon in history, it has proved to be accurate, reliable, and terminally efficient. Although somewhat heavy at 9.5 lbs., it was used by trackers in Southeast Asia, Rhodesia, and South West Africa with great success. Several versions exist, including a heavy-barrel light automatic rifle and a short-barrel model known as the "Congo."

Magnum, 10mm Magnum, .50 AE, .475 Wildey, .357, .41, and .45 Magnum rounds at a later date.

The .45 Win Mag conversion turns a rather wimpy little rifle into an ultralight, hard-hitting, low-recoiling, affordable semiautomatic suitable for SWAT and tracking operations. Assuming that you can find and buy a surplus M1 for about $200–$300, add the cost of the conversion and you have yourself an excellent multipurpose gun for about the same price as a new Ruger Mini-14.

The M14/M1A is an excellent weapon for a follow-up conducted in desert or open country where a long-range capability may be required. Although rather heavy compared with the M16, it has the range and penetration required for use in thick brush and heavily wooded areas.

.308 Winchester Rifles (7.62x51mm NATO)

Although the heaviest of the calibers under review, the standard 7.62 NATO/.308 round is still the choice of many armies worldwide.

Fabrique Nationale FN FAL

The FN FAL is probably the best-selling full-size military rifle in history. Introduced in the 1950s, it has been manufactured in many countries, is still in service worldwide, and will remain so right into the 21st century.

The FN FAL is an extremely accurate, hard-hitting weapon that incorporates range, penetration, and power into an ideal weapon for the tracker. It is a little heavy, weighing about 13 lbs. with a full 20-round magazine, but I carried one for more than 20 years and, apart from one malfunction during a firefight, never had a problem with it. As far as accuracy goes, I once saw an officer of the Rhodesian Police Antiterrorist Unit hit a guerrilla square in the jaw at a measured range of 619 yards. A lucky shot, some were heard to say, but it was an aimed shot made by a skilled marksman without the benefit of a telescopic sight.

M14/M1A and Series.

Originally designed as a replacement for the M1 Garand for the U.S. Army and Marines, the M14 was later replaced by the M16 when a light, compact weapon was needed for troops operating in the dense jungles of Viet Nam. However just because it was replaced does not relegate it to the minor leagues. The reason for replacement was doctrinal and not occasioned by any intrinsic shortcomings in either the rifle or its ammunition. Sad to say, the military hierarchy decided that the average American soldier could not shoot worth a damn, so it replaced the supremely accurate semiauto M14 with the spray-and-pray M16, a move vigorously resisted by the Marine Corps, a bunch of folks who really know how to shoot.

Gas operated, the M14 fires from a 20-round magazine and is exceptionally accurate out to 600 yards in the right hands. This weapon is recommended for trackers of a larger stature who are able to cope with the increase in weight of rifle and ammunition (200 rounds weigh 20 lbs.!).

Surplus stocks are expensive and not readily available, but a version known as the M1A is currently manufactured by Springfield Arms. The Chinese armament companies Norinco and Poly Tech also market several variations in the United States that are almost as good as the real thing. The M14SA and a shorter version known as the SAC are sold by Federal Ordnance.

* * * * *

Bearing in mind weight, firepower, ammunition capacity/weight, reliability, and handling characteristics, the most suitable rifle for trackers, in my view, is a compromise between the light/fast .223 and the heavy/hard-hitting 7.62mm NATO. This places the Soviet 7.62x39mm in the place of honor. The two available weapons in that caliber, apart from the AK-47 and the SKS, are the Olympic Arms AR-15 Hunter and the Ruger Mini-30. If cost is a factor, then the Mini-30 at around $500 has a decided edge over the Hunter, which retails for well over $1,000 at current prices. When pushed as to my choice of rifle for the tracker, I have to be honest and say I would be very happy to carry the Mini-30.

The above recommendations are mine, and I take full responsibility for them. After 20 years of carrying the reliable FN FAL in action, I know without a doubt that had I, my unit, and my army been armed with the Hunter or a Mini-30, we would have achieved a better kill rate than we did. I also understand that some police officers have no choice in the matter but to use those weapons officially sanctioned by their departments. The bottom line is simply this: whatever you have, master it and make the most of it. Make it work for you.

That reminds me of a story. During the war years in Rhodesia, a young white soldier on duty on the wall of the mighty Kariba Dam got involved in conversation with a black Zambian sentry patrolling his side of the line defining the international border. After exchanging pleasantries for a few minutes, the Zambian soldier asked if he could have a look at the FN FAL carried by the Rhodesian. They exchanged weapons, and the Zambian carefully examined the FN for several minutes. Returning it to the Rhodesian, he took back his Yugoslavian version of the Soviet SKS and looked at it with disgust. After several seconds he said to the white soldier, "The Zambian government gives us shit," and without another word, tossed the weapon over the wall to crash onto the rocks 400 feet below. He turned back toward the Zambian side and walked off slowly, mumbling unhappily to himself.

Shotguns

Despite several intrinsic drawbacks, shotguns can be ideal weapons for trackers, especially in areas where the brush is thick such as in the coastal areas of California, where visibility in the chaparral-clad hillsides is limited to only a few yards, or the rain forests of the Pacific Northwest, where the vegetation is similar to that in Asian jungles. In these conditions the choice of armament is critical, because trackers need to squeeze as many good features as possible from their weapons. In dense brush, the assault rifle in both .223 and 7.62x39 calibers is much less effective than the shotgun, which itself is hamstrung by its relatively short range and the weight of its ammunition. However, by understanding the abilities and limitations of fighting shotguns, they can be used to great advantage in situations where assault rifles are rendered virtually ineffectual.

Range

A shotgun's inability to place effective neutralizing hits on a human target at over 25 yards has been the most limiting factor in what is otherwise an excellent weapon for trackers. Chokes do have the ability to reduce pellet dispersion but do not increase range to any significant degree. The use of solid slugs in rifled barrels, however, increases the effective range up to four times, although the spread effect of the pellet pattern is lost.

Where the use of solid slugs is most advantageous is when the brush is dense and visibility limited. The heavy projectile carries sufficient mass and energy to crash its way through the brush without being deflected, as would be the case with a light, fast bullet like the .223. Therefore a compromise must be sought: a shell that combines brush-busting power, good accuracy potential, and a better than normal range. Such combinations are available, and probably the best shell to achieve these requirements is the Genco 12-gauge Duplex, which is a combination of a 330-grain lead slug and nine #3 buckshot pellets totaling 504 grains, which exits a Remington Model 870 with 20-inch barrel at 1,175 fps. The great advantage of this combination is that it is capable of both penetration and shot dispersion at an effective range of 100 yards, which is as good as you can expect to get with a shotgun. Available directly from Genco, these effective and versatile shells are highly recommended for the combat shotgun when used in the tracking role.

Capacity

Limited shell capacity is the major weakness of the shotgun as a combat weapon. In a firefight, the downtime spent reloading could well make the difference between victory or defeat.

For a period of time during the Rhodesian bush war, I was the second in command of 1 Commando, Rhodesian Light Infantry (RLI), which played a major role in "culling" a large number of our adversaries. Our mission requirement at that time was to provide a helicopter-borne fireforce on instant standby to react to any sightings or contact with terrorists active in our area of operations.

Early one morning, a radio report was received that a police patrol was following fresh tracks of about 30 to 40 terrorists some 40 miles from our base. Without further ado we lifted off. My fireforce that day consisted of three sticks of four men each uplifted by an Alouette helicopter with a 20mm gunship in support.

Within a few minutes we were circling overhead. Taking over from the police patrol, I immediately dropped my best teams onto the tracks. To cut a long story short, the eager and aggressive trackers, running on the spoor, were ambushed by the terrorist group, resulting in one tracker killed and three wounded. The bush was exceptionally thick, and a scattered but vicious firefight ensued. We spent most of the day painstakingly searching out and annihilating the gang members one by one.

Sgt. Peter White, an exceptionally competent soldier and decorated tracker who carried a Browning semiauto shotgun loaded with 00 buck, was carefully moving with his stick of four through the contact area when they were fired on by five members of the gang concealed under the dense undergrowth. Sgt. Richie Smith, a member of the Tracker Combat Unit, and an RLI trooper were both hit and killed instantly in the burst of automatic AK-47 fire. White returned fire with his six-shot repeater, downing all five terrorists. Prior to moving forward to check out the bodies, he kneeled down to reload his empty Browning. At that moment, one of the terrorists who was playing dead raised the barrel of his AK-47 and fired a single shot, striking White in the neck and killing him instantly.

The last remaining trooper of the decimated stick crawled forward, retrieved Sergeant White's radio, and informed me of the situation. It took several minutes to get to the contact area, but when I arrived I found next to White's body his empty shotgun and a handful of shells he had been loading when he was shot. I'll never know whether Sergeant White was aware that the terrorist was still alive and capable of fighting on, but the fact of the matter was that the shotgun was empty and White had to stop and reload it. Although it is easy to be wise after the event, he would have been better off in this case with his 30-round FN FAL. We lost three excellent trackers that day, the only time during the entire 15 years of war that three trackers were killed on the same day in the same action.

It is to Peter White, Richie Smith, and Doug Cookson, the other tracker killed that day, that this book in part is dedicated.

To complete the story, I circled the terrorist bodies lying on the ground and noticed that four of them were obviously dead, but one, lying face down over his AK, was still shiny with sweat. As I placed the barrel of my FN on his back, he raised his head, lifted his hands, and said, "Surrender, boss." Yeah, right!

Most high-capacity tubular magazine shotguns available today hold either seven or eight shells, which under normal circumstances are more than enough, particularly if tracking only one fugitive or a small group. A 00 buck shotshell contains nine .33-caliber pellets and if fired rapidly from an eight-shot magazine produces a formidable barrage of 72 pellets with machine gun-like efficacy. Therefore, if you are going to use a shotgun, fit it with the longest extension tube you can get your hands on and keep an additional shell in the chamber. Some guns chambered for the 2 3/4-inch shell can hold 10 when thus modified.

For rapid reloading of the shotgun, a Speed-Feed stock, where two additional shells are housed in a compartment on either side of the butt, is a big advantage. It places four shells right at your fingertips for rapid reloading. Several manufacturers produce clip-type ammunition carriers that attach to the side of the receiver and facilitate rapid reloading, especially if you have been trained in the "shoot one-load one" principle. Tac Star of Arizona manufactures the Side Saddle, which holds six shells right where you need them on the side of the gun. As each shell is fired, it can be replaced immediately with one from the conveniently situated carrier.

A new innovation that has just come on the market is a six-round tubular magazine that loads into a specially converted shotgun in the same way a fresh magazine loads into a rifle. Known as the Interchangeable Tubular Magazine (ITM), it is manufactured by New England Shotgun Technologies, Inc., of Norwalk, Connecticut. Designed for use on the Mossberg 500 Persuader, Remington 870, and Winchester Model 1300, it enables the operator to remove an empty ITM and reload with a full six-shot magazine in less than three seconds. Made of either steel or high-impact plastic, the system adds a new dimension of serious firepower to the shotgun-armed tracker.

Several magazine-fed shotguns have come on the market recently. One with a seven-shot capacity, manufactured in South Africa and known as the MAG 7, is available but has not been tested by the writer. If it functions as well as other South African shotguns, such as the 12-shot Striker, maybe it's worth having a look at it—if you can find one.

Weight

Most traditionally styled semiauto and pump shotguns weigh in fully loaded at between seven and nine lbs., which is acceptable for tracking use. There has been recent developments in the design and development of extra high-capacity shotguns such as the South African-designed Penn Arms Striker 12, which holds 12 shells in a revolver-style drum magazine. It is an excellent weapon for military or SWAT operations or for vehicular use, but the Striker is too heavy and unwieldy to be taken on a long follow-up.

Handling Characteristics

A shotgun used in the tracking role is likely to be taken into areas of thick brush, which mandates that it should be compact and maneuverable. Therefore barrel length should be reasonably short but not so short that the ammunition capacity in the tubular magazine is reduced. Most tactical shotguns have barrels between 18 and 20 inches in length and still retain a shell capacity of six to eight rounds, permitting fast, instinctive handling and aimed fire. Any barrel longer than this would restrict ease of handling.

The standard bead sights fitted on most out-of-the box shotguns are far from satisfactory for tracking. Much more suitable are the ghost ring sights that are standard on some models of the Mossberg 500/590 series or the Trac Loc sight system offered by Scatter Gun Technologies of Nashville, Tennessee. With ghost ring sights mounted, the rifled barrel shotgun is able to place solid slugs into a six-inch ring at 100 yards with monotonous regularity.

Pump or Semiauto

It is largely a matter of individual choice whether one chooses a semiauto or pump-action shotgun. Both have their strong points and drawbacks. Anybody who has seen the FBI video of the Miami shoot-out must have agonized, as I did, to watch a seriously wounded Agent Ed Morales painfully reload his Remington pump action with one good hand by holding the forestock and pushing the butt against the ground in the middle of a firefight. That he did so is a tribute to Morales' dedication, bravery, and training, but if the gunman Platt had been less seriously wounded than he was, there is no doubt that Agent Morales would have died by the hand of a man who had already killed several FBI agents. Hindsight is an exact science, but had Morales been armed with a self-loading shotgun, he would have been able to neutralize bank robber Platt and his sidekick Mather quicker than he did.

From a tactical perspective, a semiauto shotgun is much easier to handle in the prone position because it is difficult to cycle a pump action while lying down unless you roll over onto your back, which is suicidal should you lose visual contact with your assailant.

Although the recycling time is faster with a semiauto, there is no real difference in the handling and power characteristics of both systems if the same ammunition is used.

Power

A principal characteristic of an unchoked shotgun is the ability to handle without modification a variety of cartridge loads. A shotgun can fire buckshot pellets of all sizes, solid slugs, gas cartridges, flechettes, shock rounds, less-than-lethal rubber balls, pyrotechnic rounds, chemical loads, stun bags, incendiaries, salt, rusty nails, or just about anything that will fit into the barrel. Its endless versatility and dependability has earned it an enviable reputation as the most popular firearm of all time.

Accounting for probably more than 90 percent of all shotguns produced worldwide, the ubiquitous 12-gauge offers four choices of case length: 2 1/2 inch, 2 3/4 inch, 3 inch, and a 3 1/2-inch magnum shell. Obviously the larger the shell the more power it contains. Most modern shotguns are made to chamber 2 3/4- and 3-inch shells, but an increasing number of manufacturers offer the longer chamber, which can accommodate all lesser case lengths.

Bear in mind that the magazine capacity is affected by the length of the shell selected. A 3 1/2-inch chamber gun with a 20-inch barrel holds only five magnum shells but seven of the shorter 2 3/4-inch shells. You then have a trade-off: more power/less shells or less power/more shells. As a tracker, I would go for the less power/more shells option.

Of growing interest to law enforcement is the 10-gauge magnum, which is even more powerful than the 12-gauge. Currently, only Penn Arms makes the tactical 10-gauge, 12-shot Striker, but several well-known manufacturers are considering offering their popular law enforcement models in the larger, more powerful loading. Unless you are of larger than normal stature or highly experienced in handling heavy shotgun loads, stick to the lesser gauge.

Model	Barrel Length	Empty Weight	Shell Capacity
MOSSBERG			
Special Purpose 500	18" or 20"	6 lbs.	6 or 8
Military Model 590	20"	8 lbs.	9
Ghost Ring Model	18" or 20"	8 lbs.	6 or 8
Mariner Model 500/590	18" or 20"	8 lbs.	6 or 9
WINCHESTER			
Model 1200 Defender	18"	6 lbs.	7
HIGH STANDARD/MITCHELL			
Model 9113 SAS	18" or 20"	8 lbs.	7 or 8
REMINGTON			
Model 870 (Police)	18 or 20"	6 lbs.	6
Model 870 Marine	Mag 18"	7 lbs.	8

Pump Shotgun Choices

Bearing in mind the more specific needs of a tracker, the following pump shotgun makes and models are considered suitable for the task.

Note that Scattergun Technologies of Nashville, Tennessee, specializes in converting the Remington Model 870 into what it calls the Tactical Response Concept. These superb conversions, adopted by the U.S. Border Patrol, feature the Trac Loc ghost ring sight system, which is adjustable for windage and elevation and includes a tritium insert front sight that glows in poor light. Also featured are magazine extensions, sling mounts, combat recoil pads, heavy-duty magazine springs, jumbo head safeties, and a three-way adjustable sling similar to the BEAM sling. Whereas the unmodified Model 870 is considered adequate, the Scattergun conversion puts it into the superlative class.

Although not listed above, Ithaca's M37 model, which was popular with Special Forces and recon teams during the Viet Nam conflict, is back in production. It comes in a left-hand model for the 15 percent of you who are so "dexterously challenged." Production models are available in military green with parkerized parts, making it an ideal off-the-shelf tracker's weapon. The addition of ghost ring sights would be a decided advantage.

Autoloading Shotgun Choices

The Remington 11-87 semiautomatic shotgun is already a police classic in use throughout the world. It is available in two barrel lengths, 18 and 21 inches, housing seven or eight shells respectively. The gas-operated autoloading action is very reliable and, with its low recoil, can handle any heavy buckshot or slug load. Its mil-spec parkerized metal finish and rugged synthetic stock guarantee years of maintenance-free operation. As is the case with the pump Model 870, Scattergun Technologies specializes in upgrading the 11-87. With Trac Loc ghost ring sights, tritium front post, and many other features, the upgraded 11-87 is one of the best shotguns for tracker combat operations.

If ever there were shotguns made with trackers in mind, Benelli (part of the Heckler & Koch group, now British owned) makes them. Seven different models in fact! Benelli manufactures two separate series: the semiauto M1 Super 90 and the select-fire semiauto/pump M3 Super 90 series.

M1 SUPER 90 MODELS

Model	Furniture	Barrel	Weight	Shells	Sights
Super Defense	Fixed stock, p/grip	19"	7.9 lbs.	7	rifle sights
Tactical Shotgun	Fixed stock, p/grip	18"	6.5 lbs.	7	ghost ring
Slug Gun	Regular fixed stock	19"	6.7 lbs.	7	ghost ring

M3 SUPER 90 MODELS

Model	Furniture	Barrel	Weight	Shells	Sights
Pump/Auto	Fixed stock, p/grip	19"	7.9 lbs	7 shot	optical
Pump/Auto	Folding stock p/grip	19"	7.6 lbs	7 shot	ghost rings

The M1, with its high-strength polymer stock and chrome-lined barrel, can fire five shots of any 2 3/4-inch or 3-inch magnum shell in less than one second. Several barrel options, magazine extensions, ghost ring sights, pistol grips, and regular stocks promise a variety of configurations to please any discerning operator. The unfortunate fact of life is that these guns fall under those forbidden by the 1994 assault weapons ban, so future exports into the United States are doubtful. Benellis are already expensive, so who knows what will happen to the price of any pre-ban models still available.

The M3 Super 90 series combines the semiauto feature of the M1 series with a pump-action option that can be activated quickly by turning a spring-loaded ring located at the end of the forestock. It also can be fitted with an optical sight system.

Manufacturer of the military M9 pistol adopted by the U.S. Army and countless police departments across the country, Beretta is also the maker of two excellent shotguns, both eminently suitable for the tracking role. Model 1200FP is a semiauto with a 20-inch barrel holding 6+1 shells. Fitted with adjustable sights, a matte black finish, and durable plastic furniture, the 1200FP should provide years of trouble-free service. The M3P series is Beretta's newest law enforcement shotgun that, like the Benelli M3, is convertible from semiauto to pump mode. Firing from a five-round box magazine with a folding stock, the M3P is also highly recommended for tracker use.

Submachine Guns

I am hesitant to recommend the submachine gun as a tracker's weapon for two reasons. First, the pistol-caliber bullets fired by sub guns lack velocity and range to be really effective at longer distances. Second, the light 9mm bullets (115–129 grains) lack the ability to penetrate vegetation.

This is not to say that the modern submachine gun is totally unsatisfactory in this particular role. It's just that typical sub gun calibers are marginal manstoppers at best. To give an example of its shortcomings, we have to go back to Kenya, East Africa, in the late 1950s when the Mau Mau emergency was at its peak. The Mau Mau were gangs of young men from the Kikuyu tribe who were fighting for independence from Britain using barbaric methods.

In this particular story, British soldiers had located the area where an extremely vicious gang was hidden high up in the Abadare forests on the flanks of Mount Kenya. With its thick bamboo groves and permanent wet and gloomy weather conditions, the area was not a pleasant place. Intelligently analyzing the situation, a Brit patrol lay in ambush where a track crossed a fast-flowing stream. Although the patrol had spent several days in position, cold and wet to the bone, it was nonetheless determined to close with and wipe out as many of the gang as possible.

The gang, sensing danger, had decided to vacate the area and move to a safer hiding place. Moving stealthily along the trail, senses tuned to fever pitch, they crept closer toward the ambush site. Greeted by a sudden burst of automatic fire, they fled across the creek. Several fell dead or wounded into the water as the remnants scattered toward the bamboo groves on the other side of the creek.

The gang leader, a murderous villain

Although ideal for urban operations, the 9mm submachine gun suffers from lack of range in open areas and poor penetration in thick brush. However in good hands it is accurate and can be used to put down effective suppressive fire. Its lightness and maneuverability are its best assets.

named Dedan Kimathi long sought by the authorities for his many crimes, was wearing an old British army coat made of heavy woolen cloth festooned with pots and personal belongings, as was the habit of the gang. From a range of less than 30 yards, Kimathi, running for his worthless life, was fired on by a soldier armed with a Pachett 9mm submachine carbine. Scampering down the streambed, Kimathi escaped despite being struck several times in the back by the ineffectual 129-grain full-metal-jacketed 9mm bullets.

During the after-action debrief, the soldier who had fired so accurately on the gang leader complained bitterly that he clearly saw the bullets strike but bounce off Kimathi's back. Later tests actually proved that the 9mm bullet did not even penetrate a wet army blanket hung at double thickness over a line. As a result of his "miraculous" escape, Kimathi was accorded mystical powers by the simple and superstitious Kikuyu peasants, which naturally was thoroughly encouraged by Kimathi. Claiming that he could not be killed by British bullets, he rampaged for several more years murdering and mutilating local Africans with his razor-sharp *panga* before he was finally tracked down and caught by police and eventually hanged for his crimes.

Today, major weapon manufacturers like Heckler & Koch, Colt, and Olympic Arms are offering submachine guns in new and more effective calibers such as the .40 S&W and 10mm. This has considerably "up-gunned" the wimpy 9mm. At this point, it is too early to say whether these new calibers will prove to be any more effective than the 9mm FMJ, thus making them suitable for use in the tracking role. In the meantime, if sub guns are used, the ammunition selected should be as heavy as possible—147 grains for the 9mm, 180 grains for the .40 S&W or 10mm, and 230 grains for a .45 to punch through leaves and branches and retain enough terminal energy to achieve satisfactory results.

It is interesting to note that the U.S. Army and Marine Corps do not have a submachine gun in their inventories. This is because the powers that be do not see the need or suitability of such a weapon. The Heckler & Koch MP5 is used by several special operations units for specific roles such as hostage rescue, where dangerous encounters take place at minimal ranges, but the weapon is not general issue. Until sub guns prove themselves effective in combat, they will remain a special-purpose tool for special purpose applications.

MAGAZINES

As with ammunition, always purchase good brands of magazines, preferably those made by the original manufacturer of the weapon you are using. Avoid suspect, foreign, or unknown brands, especially if they are of ultrahigh capacity or sold at bargain prices. In my experience, they always seem to let you down when you need them most. Magazines should be rotated regularly during training so that they all get equal use and should never be left loaded for long periods with springs fully compressed. Any damaged magazines must be separated immediately and marked clearly in some way showing that they are not to be used on operations.

It is always a good idea to slightly underfill magazines. For example, load 28 rounds into a 30-round mag. This way you can be sure they will seat correctly when inserted into the gun and feed smoothly without malfunction. Some brands of magazines fail to seat if they are filled to maximum capacity.

Many people seem to think that magazines need not be cleaned. This is an extremely foolish belief and one that could be hazardous to your health if allowed to continue unchecked. It is amazing how much dirt and grunge manages to find its way even into full magazines. Only regular stripping and cleaning can ensure reliable feeding and function. When cleaning magazines, always check the feed lips for damage, especially any dents or bends in the metal that can easily occur if dropped onto the open end on hard ground. Check to see if the platform is inserted correctly. In some models the platform can be fitted the wrong way, and although it may look okay, this will certainly cause a malfunction when the weapon is fired.

OPTICAL SIGHTS

When discussing firearms in the tracking role, the question of the use and suitability of optical sights always crops up. I see very little use for a telescopic sight unless the follow-up takes place in the wide-open spaces of a desert where a long-range shot is required. What has proven to be advantageous, however, is the red dot or single point sight, both of which can dramatically enhance target acquisition and shot placement in fast-moving combat situations. Correctly called an "optically occluded gunsight," you don't look through it as you would an ordinary telescopic sight. By keeping both eyes open and placing the sights in front of your eyes, a light-enhanced dot similar to a laser can be seen. Aligned with the bore axis, the dot is superimposed on the target prior to firing.

I first became aware of the single point sight in 1969 during the Rhodesian war when I was given a green dot sight manufactured by Weaver for evaluation. I used it often during tracking training, but the first time I used it in a combat situation was when I was a troop commander with the Rhodesian Light Infantry on fireforce duty.

We had been dropped in a blocking position by helicopter while an air force spotter plane fired high-explosive rockets into the general area of a terrorist meeting place sited in a heavily treed ravine. As the terrorists broke cover and ran, one of them dashed across an open patch to my front. Aiming with a slight lead, I fired rapidly as he accelerated for the cover and safety of a wooded gully. I don't know who eventually hit him; all of my four-man stick was firing madly, but I do remember thinking at the time, "Wow, this is great. Every soldier should have one!" Not only was I able to track the running man but, having both eyes open, I was fully aware of what was going on to my front. When a second guerrilla broke cover in the tracks of the first, I was able to immediately swing my weapon over and engage him with effective aimed shots, dropping him before he had gone more than a few yards. After the dust and excitement had died down, we examined the scene and were surprised to find not two but three bodies lying in the open patch.

There are two types of optically occluded gun sights. The first type (and the least expensive) gathers light through a dome-shaped extension to the sight tube that illuminates the dot. The drawback of this type is that it can only be used in daylight unless an expedient light source is attached for night use. Armson of South Africa makes a simple but effective version called the OEG that sells for $75.

The second type is more sophisticated and powered by a watch-sized lithium battery so that it can be used both day and night. I currently use an Aimpoint 3000 red dot sight on my M1 carbine. It has a 10-setting variable light switch for use under any light conditions.

There are several battery-powered red dot sights on the market. Aimpoint has a special deal for law enforcement that drops the price of the company's 3000 model from $232 to $186. Mounts are available for all weapon makes and models, from handguns to submachine guns to shotguns to rifles. After being available for nearly 30 years, several Western armies, including the U.S. Army, are now looking to equip their soldiers with optical sights, and the red dot system is in the forefront of consideration. If you are looking for a definite combat edge, the red dot sight will give it to you.

SLINGS

Generally speaking, a sling should never be used in combat situations, and particularly not in the tracking role. Your primary weapon should be in your hands ready for instant action if required and not slung Hollywood fashion over your shoulder. There are, however, several new tactical slings designed for SWAT operations that enable fast shouldering of the weapon and secure carry if a secondary weapon is used. The BP16 sling made by JFS of Salem, Oregon, and adopted by the Marine Corps is a fast-action assault sling with considerable promise. Several other slings made of elasticated material that permits fast action and handling are currently available too. A word of warning here: if you choose to purchase one of these new newfangled things, *practice, practice, practice* until you are completely certain there will be no snags or foul-ups when you have to use your weapon in action.

In the tracking role, a long gun can also double as a tracking stick and be ready for instant action if held where it ought to be—in the hands. Trackers never know whether or when they are going to be ambushed, so, like the Boy Scouts, they should always be prepared.

LOAD-BEARING EQUIPMENT

Thank heaven the old days of uncomfortable, impractical, painful, and dysfunctional load-bearing equipment are over. Modern synthetic materials and plastics with exceptional strength, durability, and low weight, along with a better understanding of ergonomics and design, have revolutionized this important but neglected area of a soldier's (and policeman's) essential equipment. One only has to compare backpacks of just 10 years ago with the sleek, trim, functional models available today to see what I mean.

The concept of "form fits function" dictates design for most things these days, a fact that has not been lost on both producers and users of tactical equipment. The LBE designs of Eagle Industries are excellent examples of the truth of this principle. Their modular vest system, Model LBV-M, can be fitted with a variety of moveable pouches attached to the basic vest using Fastex Twist Lock fasteners. Suitable pouches and panels can be fitted in whatever configuration is desired according to the weapon, ammunition, equipment, and mission requirements of the team or each individual member. Available in either olive drab or woodland camouflage and manufactured from high-durability Cordura, all load-bearing points are triple stitched for strength and durability. Pouches are available for magazines of all major weapon systems as well as for shot shells, first aid kits, and radios. Belt loops are built in if a web belt is to be worn.

Eagle Industries' LBV-D model also offers enormous flexibility and economy with different pouches fitted on Velcro panels according to stature and operational or weapon requirements. Additional fittings are mounted to secure the modular pouches in place to prevent slipping and sagging. Like the LBV-M, a pistol belt can be attached to hold additional gear. One size fits all with adjustments for height and girth.

The U.S. Cavalry Store stocks the excellent Harris Assault Vest designed for military special operations. This rig features four slanted magazine pouches, two side utility pouches, and four smaller pouches for compass and other essential small gear. A neat backpack clips to the shoulder harness, and D rings abound for attaching other gear. It is available in camouflage.

Whichever equipment you purchase, it is strongly suggested that you give it a real test run in the woods for at least a full day of strenuous activity to work out all the bugs and adjust it correctly and comfortably to your body shape and size. Ill-fitting gear can cause you needless pain and suffering if you fail to test-wear it and get accustomed to its peculiarities.

If your budget does not stretch to an expensive modular system, there is nothing wrong with the good old G.I. suspenders, belts, pouches, and canteens available from your local military surplus store. When fitted correctly and well worn in, this rig will give you years of good service.

BOOTS AND FOOTWEAR

Footwear technology has made enormous strides (no pun intended) in the past few years. Fueled by the sportswear and fashion industries, exotic materials such as Gore-Tex,

Cambrelle, Thinsulate, Cordura, and polypropylene have revolutionized the industry, resulting in the best range of footwear ever for trackers. When you choose a boot, take into account the following requirements:

- Comfortable to wear for up to 48 hours at a stretch
- Rugged but lightweight to prevent fatigue
- Waterproofing to ensure dry, abrasion-free wear
- High tops to keep out stones and debris
- Ruggedness and durability to withstand abrasion from rocks
- Easy on/easy off, preferably with hooks or similar fastenings
- Suitable sole pattern with excellent traction qualities (see below)
- Good ankle support to prevent injuries

Sole Patterns

There are two schools of thought on the choice of a sole pattern suitable for trackers: cleated or plain sole. As far as tracking is concerned, the heavy lug or Vibram pattern so popular with many police and military folks is undesirable for two reasons. First, a heavy, deep-cleated sole is usually attached to a heavy boot, and second, it tends to pick up a lot of mud or clay, which can bring the weight of the boot up to five lbs. Imagine that each boot holds two lbs. of useless mud, and there are 2,000 paces to the mile, and you begin to see how much extra energy you will have to expend. If a boot with a cleated sole is selected, the sole should be either a shallow design like Hi-Tech or the military Panama style, which is specially designed to shed excess mud. Most SWAT-type boots made by Danner, Rocky, Hi-Tec, or Rainier are suitable and are found in most police supply stores.

The other school of thought, to which I belong, prefers a completely plain, unpatterned sole with rounded edges so that no recognizable spoor is left behind. Rhodesian and South African trackers were issued a high canvas boot with a totally plain, rounded sole that was excellent in the dry, sandy conditions of the African veldt, but I fear they would not be suitable for the wet, muddy conditions found in the United States. "Slip-sliding away" would be the order of the day. My current boot for dry, sandy areas is a high-top leather model that used to have a shallow bar pattern, which I removed with a wire wheel fitted onto a bench grinder so that no pattern would show on soft ground. I suppose, however, that the first time I slip on a wet patch of grass or mud and land with a bang on my butt, I'll change to something more gripping.

If it is at all possible, the entire team should wear the same pattern to avoid possible confusion when following a spoor. The U.S. military issues a suitable jungle boot fitted with the standard Panama sole. It is an ideal choice, being light, strong, cool, and durable, but it is not waterproof.

Boots are an intensely personal item, so much so that the Rhodesian Army turned a blind eye to a bizarre selection of footwear, from rope soles to bare feet. The prevailing policy was, "What is comfortable for you is comfortable for the army too." Perhaps it is well to follow that policy and wear what suits you best. Put another way, don't judge a man's taste in footwear until you have walked a mile on his feet!

Socks

Just as important as footwear is to a tracker is his choice of socks. Poorly designed socks or those made from an unsuitable material can be disastrous for a tracker on a long follow-up and may put him out of action unnecessarily. Several brands of polypropylene socks are available that wick moisture away from the feet to prevent "hot spots" and blisters from forming. Thorlo's Mountain Climbing or Trekking socks and Wigwam's Ultimax hiking socks are recommended. Better yet, if you can get away with it, is not to wear socks at all. Increased comfort can be obtained by inserting shock-absorbing inner soles or heels, and military-issue mesh inserts are first rate for keeping feet cool under hot conditions.

UNIFORMS

Military-style camouflage uniforms of the type worn by many SWAT teams are perfectly satisfactory for tracking use. The most popular is the woodland camouflage pattern, followed by the Viet Nam-era tiger stripe pattern. Avoid those with a predominance of black (no more than about 15 percent of the total color), and under no circumstances should urban camouflage or black SWAT uniforms be used in the woods. The idea of a camouflage uniform is to break up the outline of the body and blend into the background. If a uniform that is intended for daytime use has a large percentage of black, the body shape is emphasized and will stand out against lighter backgrounds. If you take a look into the woods, you will not see anything that is black. Even shadows, which we assume are black, are actually various shades of gray. Generally speaking, faded camouflage is better than fresh, contrasting colors, and if there is no choice, olive drab jungle fatigues are far better than wearing the wrong color and type of camouflage.

There are three different materials used to manufacture uniforms, each one having its own characteristics and specific uses. They are 100 percent cotton, cotton-poly blend, and ripstop cotton. The cottons are definitely cooler in warm climates, and the ripstop is more durable and more suitable overall. The cotton-poly blend is usually denser and a lot hotter and more uncomfortable to wear in warmer climates.

As a tracker, it is likely that you will operate the year round, so you must choose your uniforms according to the predominant colors of the vegetation and terrain in which you expect to operate. If this is a problem, then chose a compromise pattern that fits best into a variety of color backgrounds. For an all-purpose generic pattern, ripstop woodland is about the best, especially if it is faded to about 25 percent of the original color tones. A visit to your local surplus store or flea market should get you what you need at a reasonable price.

The hunting and camouflage clothing industry has exploded in the past few years, with new patterns and colors appearing almost weekly. A lot of the designs are based on scientific principles and are very effective, but some seem to have been concocted in the chimpanzee cage at the local zoo. One excellent pattern receiving growing attention is known as Timber Ghost, which at the time of this writing is available only in shades of brown, tan, and dark gray. This pattern appears to be based on a World War II German design that was issued on a limited basis to paratrooper units. The way the colors are applied creates a window effect, giving the appearance of being able to see "through" the garment. It is particularly effective in light- and shade-dappled areas.

A boonie-style hat is preferable to a police-style baseball cap because the latter's distinctive shape is conspicuous in a woodland setting. The shape of the human head is a dead giveaway, and every effort should be made to break up the outline. The boonie hat with attached loops lends itself to good camouflage techniques.

RAIN GEAR

The standard U.S. military-issue poncho is a good, inexpensive rain protector that also doubles as a ground sheet or shelter if required. The only problem with the poncho is that it can be cumbersome and constricting, especially in thick brush. I have never engaged in a firefight wearing a poncho in the rain, but I imagine that it would prove to be a hindrance when wanting to extract a fresh magazine, for example. Better than a poncho is the new G.I. rainsuit made from a lightweight double-coated polyurethane that permits your gear to be worn over the top, allowing instant accessibility. In a emergency you could always use a full-size trash bag with holes cut out for head and arms worn under your outer garments. This will also form a vapor barrier that helps retain body heat.

Whatever garment you buy, be sure it is efficient. It is amazing just how much operational enthusiasm can be lost when soldiers or officers are wet and cold to the bone. Hypothermia is an everpresent threat and should never be taken lightly. Seek a compromise between lightness and warmth, especially if the weather forecast predicts inclement weather.

KNIVES

A good knife is more than just an essential item, it is a friend. This is true not only for a tracker but for anyone who ventures into the backwoods. If you don't believe this, go into the woods for several days without one and you will soon see just how many things you cannot do because you don't have a knife with you. When it comes to knives, there are three choices: fixed blade, folding models, and multiple use/gadget knives.

Fixed Blades

A good, inexpensive fixed-blade knife is the workhorse of World War II, the venerable Ka-Bar. Having seen service in every corner of the world, it is still issued by some armies and is available in most every surplus store. For a $30 knife, that should say a lot for its versatility and durability, and if it should ever get lost its not expensive to replace. When considering a knife, I always imagine exactly what it could be used for, from slashing a path through thick brush to chopping poles for an expedient litter to trimming one's fingernails. The Ka-Bar has always worked for me.

An excellent range of innovative knives called Spec Plus, manufactured by the Ontario Knife Company of Franklinville, New York, have recently come onto the market. Described as the next generation of military knives, they range from pocket folders to combat/survival knives to several styles of Bowies and machetes. Ruggedly made of carbon steel with a black epoxy, powder-coated finish and nonslip Kraton handles, they are absolutely ideal for any tactical environment. While instructing tracking classes, I regularly carry Ontario's Model SP8 Survival Machete, probably the most useful and versatile bush tool I have ever encountered. With a squared-off 10-inch blade and saw back, I have used it for everything from digging holes to chopping firewood to opening wooden crates. It comes with a clever fast-opening Cordura and leather sheath, and it cost me less than $50. Check out Spec Plus blades: you will be impressed.

Folding Knives

Spyderco manufacturers an excellent range of folding knives that are sturdy and dependable and that can be opened with one hand. I carry one with me everywhere and am constantly amazed just how many times in a day I find use for it, from sharpening pencils to opening letters to removing a sliver. Spyderco knives are first-rate, well-designed tools available from all good police supply stores and military mail-order catalogs at reasonable prices. All folding knives made by Blackjack are also highly recommended; they are made of first-rate steel and keep an incredibly sharp edge.

Multiple-Use Knives

Multiple-gadget knives have come a long way from the old Swiss Army knife, which is good but does not fill the exacting requirements for a rugged backwoods tool. (Although to be perfectly honest, I once witnessed the amputation of a severely injured leg with a Swiss Army Champ model. A RENAMO (the mostly South African-backed political-military force that opposed FRELIMO in, mostly, the 1980s) guerrilla in Mozambique was struck in the leg by a rifle grenade, blowing it completely away. He subsequently walked for over 20 miles to get medical assistance. The entire operation was done with the knife because we had nothing else to use. It did everything from sawing through the bone to cutting a flap of skin to covering the end of the stump. It testifies to both the surgeon and the knife that the man survived his ordeal and was last seen hobbling back toward his home on a makeshift crutch.)

All the Leatherman models, Gerber's Multiplier, and the Toolclip from SOG are excellent examples of functional, useful tools that should be carried by a resourceful tracker. There is practically no end to the uses for these tools, and anyone who ventures out of doors should have one tucked away somewhere.

* * * * *

My personal choices for good all-round bush blades are the Ka-Bar and a black, regular-sized Leatherman tool. I have yet to find a better combination. For bush expeditions I always take along my Spec Plus machete too.

A word of warning here. Knives are an intensely personal item for most people, and the United States has some of the best custom knife makers in the world. Being handmade, they generally cost big bucks, so before you take an expensive custom knife into the woods, consider the financial impact if you should lose it. I have lost several good knives over the years, and if anything is likely to get misplaced in the wilds, the knife is the leading candidate. There is a wide selection of excellent but reasonably priced knives available. Rather lose one of those than risk a more expensive one.

COMPASSES

A compass is an essential tool for every member of a tracking team. Every tracker should always wear a wrist compass so he can maintain direction, particularly in thick brush or heavily wooded terrain where disorientation is always a possibility. As a tracker, I always found a wrist compass invaluable when exiting a helicopter to obtain a rapid sense of direction. (No matter how good you think you are at estimating direction, there is nothing like a helicopter descent to prove you wrong.) A good wrist compass should cost no more than a few dollars from your local surplus or sporting goods store, although there are some very expensive models available.

We have discussed binoculars and watches that display direction, but it is just as well that at least one person in the team have a good, accurate, hand-held compass in case batteries run out or delicate technology goes awry. Compasses get lost easily, so attach yours securely to your gear with paracord.

Prismatic compasses are expensive but worth their weight in gold. Should your budget be more modest, then I suggest you purchase a mil-spec Lensatic compass for about $40. This is an adequate compass and has served the military well for many years. It can shoot an azimuth, orient a map, and be used as a marching compass if necessary. The U.S. Cavalry Store has a good selection ranging in price from only a few dollars to the super-deluxe M2 tritium model at around $160. Brunton of Riverton, Wyoming, manufacturer of the M2, has a whole range of compasses from wrist models to really sophisticated types suitable for basic surveying and map-making purposes. Silva, the Swedish manufacturer, also sells a good line of accurate and dependable compasses ranging from a few dollars to around $50.

Do not purchase a cheap foreign compass, especially those made in the Far East. There is a strange law in the backwoods that cheap equipment breaks down the moment you need it most. If you want good service, get the best you can afford and then some. It's worth it when your life is at stake.

MAP READING RESOURCES

If you or your team need map reading training, there are several books and videos available that give an excellent overview of the basics. (Brunton sells an excellent video entitled *The ABCs of Map and Compass* for $34.95.) This, however, is not enough to master the subject. You *must* get out into the countryside, preferably with a good navigator, and learn it for yourself. Like riding a bicycle, you will never lose map-reading skills once you have mastered them, but only experience on the ground will lead to mastery.